The Energy Solution Revolution

*A socio-political journey through the tangled
world of free and clean energy,
its promise, its suppression and its logical
necessity for our survival*

Brian O'Leary, Ph.D.
www.brianoleary.com

Second Edition, Copyright 2009, Brian O'Leary

For permissions, or serializations, condensations, adaptations, or for our catalog of other publications, write the Publisher at the address below.

Montesueños Press

P.O. Box 258
Loja, Ecuador
drbrianoleary@gmail.com
www.brianoleary.com

Published By:

Bridger House Publishers, Inc.

PO Box 599, Hayden ID, 83835, 1-800-729-4131

ISBN:
978-0-9799176-4-6

Cover images copyright Brian O'Leary
Typesetting/Design: Julie Melton, The Right Type Graphics

Manufactured in the United States of America using digital manufacturing which has significant environmental advantages over traditional offset printing. One-book-at-a-time manufacturing eliminates supply chain waste, reduces carbon offsets and conserves valuable natural resources.
10 9 8 7 6 5 4 3 2

Contents

In memory of my mother Mary Mabel O'Leary, who gave me life and so much more. Among her last words were "I am nothing." May the rest of us learn and experience that nothing is something.

and

In memory of Meredith's mother Barbara Miller who gave her life and so much more. Among Barbara's last words were, "I feel that half of me is no longer here." May the rest of us learn about the void we must all enter one day.

and

for the unsung and courageous free energy researchers who know from their experiments that nothing is something, and have sacrificed their lives and toiled for decades to bring to all of us miracles of change, if only we look and act.

Foreword

In our upside down modern world, where coal has become 'clean' and nuclear power 'green', what becomes of those who truly seek to improve the lot of society, those with the imagination and the intellect to think of sustainable brave new worlds of energy independence?

The following short story might provide some insight to that question, and to Dr. O'Leary's central premise in this courageous and much needed book:

The Last Wizard

A ripple of attention followed Mark Twain as he strode through the Palm Room of the Waldorf Astoria, tossing off quips to members of New York's social elite, the '400'. But much as he enjoyed this attention in the twilight of his life, Twain's gaze led him to a pool of calm at the side of the dining room.

Resplendent in a white dinner suit, Twain took a seat opposite a middle-aged, dark-haired gentleman of European descent. "Well, some things never change!" He gestured impishly at the eighteen napkins on the table, stacked neatly in three piles of six. "When everyone else is ready for desert, you'll still be polishing the cutlery and crystal!"

The man nodded to Twain. "I was just observing my new bladeless turbine," he said, his accent slight.

"In your head?" Twain laughed, knowing full well that his inventor friend did just that, preferring to design and test new inventions in his marvellous mind prior to producing final working types.

The inventor elaborated, "It currently weighs less than ten pounds, and as of this evening is developing thirty horsepower." He made a dismissive gesture. "I can do better." The maitre d' delivered his dinner, prepared off-menu as per specific instructions telephoned hours before.

Twain lit a cigar while his companion scanned the meal set before him. "Well my friend," he drawled, "what's the verdict?"

"Twenty-four point four cubic inches," he answered with certainty. He allowed a rare genuine smile. "Entirely sufficient. We must leave room for dessert, Mark."

Twain blew a smoke ring. "Any luck with your wireless torpedoes?"

The inventor's mouth twitched. "The military have no interest in my teleautomatics." He shrugged. "Perhaps it is for the best: it may be that I have not the right to invent such instruments of war. But the moneys would have been put to good use at my Wardenclyffe tower." His voice became distant. "I offered J.P. Morgan a monopoly on radio broadcasting... 'you will be able to broadcast all wavelengths from a single station, across any ocean,' I told him. 'You will have a complete monopoly; the whole world will be listening...' Now, he shuns me, when I most need his backing."

Twain placed a hand over his friend's. "Yes, but you forgot to mention your other purpose, to transmit electrical energy without wires to any place on the globe. A banker has no interest in beaming electricity to aboriginals; not when it can't be metered."

The inventor's eyes twinkled momentarily. "A regrettable lapse on my part," he agreed. His tone sharpened. "It is such a simple feat of engineering, Mark, but the blind and fainthearted cannot see past their own immediate interests."

A silence settled over the table. The inventor polished his knife and fork again and began at his meal.

Mark leaned back in his chair, gave his cigar a puff, and blew another smoke ring to the ceiling far above. He'd been born the same year Halley's comet last appeared, and had long maintained a belief that he should exit life's stage when it next crossed the heavens. That would be in 1910, allowing him just five years to enjoy his fame and recent prosperity. There would be plenty of time for melancholy when he was dead! His thoughts drifted. In the old days—the early nineties—his inventor friend had entertained Twain and other curious gentry at his private loft laboratory at nearby 33-35 South Fifth Avenue. What thrills they had seen in those times, when in the middle of the night the inventor would throw a switch, turning darkness into a wizard's lair eerily lit by tube lighting with no electrical connections. At the mere click of his fingers, the inventor created fireballs that left no marks on his hands as he carried the red flames about the laboratory, casting shadows on the lumps of strange apparatus scattered throughout. Then there were the blurred photographic plates produced by the effect of what the inventor referred to as "invisible light," and the tiny pocket-sized mechanical oscillator mechanism that, when once attached to an iron beam that ran from his laboratory to the building's foundations below, had created tremors in the surrounding neighbourhood streets, shattering windows and shaking buildings. When the local police—who were ever suspicious of his endeavours—arrived, the inventor had already smashed the device with a sledgehammer. He apologised that they were too late to witness his little experiment, but were welcome to visit again later that evening. But now, he had told them, they had to leave as he had important work to do...

In those days the inventor had been the happiest man in Manhattan. Yes, the inventor had had wealth, but he had given little thought to it so long as he could maintain himself at the Waldorf and immerse himself in his research. Indeed, after giving four fantastic lectures in America and Europe, the inventor had become the world's most renowned scientist. He had become a celebrity, a scientist showman the likes of which had never been seen before. But when the man who had bathed himself in hundreds of thousands of volts of electricity on stage, and regularly sent lightning bolts to the sky in Colorado, claimed that he had listened to signals from space—possibly Mars—the tide began to turn, and swiftly. It was too much for a stolid scientific establishment keen to seize upon any opportunity to claw back public attention from this unschooled prodigy, whose work they could barely comprehend.

The inventor finished eating. Breaking the silence, "Mark, we must wean the world from oil, wood and coal."
Twain roused from his memories. "Windmills on every house, eh?"

No. Although wind and solar power would be sufficient first steps, better than crawling as we do now in muck and soot created from current methods of production." The inventor smiled again, but sadly: "We cannot dig and burn forever, Mark; the Earth has limits." He stared past Twain, into a distance only he could see, perhaps already constructing new working marvels in his mind. "It is the only practical solution: you see, Mark, there is no need to transmit power at all. We must attach our motors to nature's wheel. One day, when humanity grows out of its pedantry, stupidity and ignorance, machinery will be driven by a power obtainable at any point of the universe."

Twain bit down on his cigar. He knew people very well, having made a successful living writing about them. He cast a glance about the Palm Room, at the members of the influential 400 sipping brandy and port as they discussed who was saying what about whom and whose fortunes would be made or destroyed the following day: frolicking in muck was what they did all too well...

***

Nikola Tesla, father of radio, AC motors and power distribution and other marvels, never since replicated, died penniless in 1943. Despite being a US citizen, the FBI released his estate to the Alien Property Office, and on the advice of the War Department—who had hitherto shown no interest in his proposed particle beam or "teleforce" weapon—Tesla's papers were classified "Top Secret."

It is fitting to say that Tesla made the world you see today possible.

Tesla might have added, "I would have given you tomorrow's world, too…"

In *The Energy Solution Revolution,* Dr. O'Leary asks us what sort of world would we like to inherit in 2050—a bleakly desolate totalitarian world, or a global green democracy, where our leaders are always visible, always accountable. The choice really is ours to make, and for our children and those to come, we must choose.

I hope that Dr. O'Leary and the New Wizards of the 21st century fare better in their quest than did Tesla.

Dr. Shaun A Saunders
Psychologist & Author of *Mallcity 14*
Sept. 2, 2008
Newcastle, Australia

Introduction

"The richest 400 Americans...that's right, *just four hundred people*...own MORE than the bottom 150 million Americans combined. *400 rich Americans* have got more stashed away than half the entire country! Their combined net worth is $1.6 trillion. During the eight years of the Bush Administration, their wealth has increased by nearly $700 billion—the same amount that they are now demanding we give to them for the 'bailout'."

– Michael Moore, www.michaelmoore.com,

Oct. 2, 2008

One century ago, J.P. Morgan of the infamous "400" social-financial elites denied Nikola Tesla his funding for developing free energy. Morgan's monopoly of copper mines compelled him instead to build vast now-familiar landscape-littering power grids of copper wire that ironically used a Tesla invention: alternating current. What is it about today's "400" club of billionaires that echo what happened to Tesla? For, as we shall see, Tesla is not the last wizard. Hundreds more are coming forward if we only give them a chance. But, as we shall also see, history keeps repeating itself and we never seem to learn. Out of a greedy self-interest, our rulers keep acting the same way as Morgan did. But the stakes are much higher now because of the pending death of nature and civilization. We will have to end the biocidal plutocracy responsible for the mess.

This book is an urgent call for an energy revolution that would give us a quantum leap in environmental friendliness. Some parts are an updated compilation of essays written over the past several years that explore the prospects and implications of a future world of radically clean, cheap and decentralized energy. Does this possibility really exist or is it a scientific mirage? If it does exist, why has it been so suppressed for over a century since Nikola Tesla and others were inventing these things? Why do mainstream scientists, progressives and environmentalists totally ignore this possibility, even though it might save us all? What can we do to avoid the abuse or overuse of these technologies? Most importantly, how can we implement free energy when so very few of us are listening? We explore these questions in this book.

The progression of chapters reflects my own progressive thirst for the truth in the face of my increasing outrage at the violent suppressions of sustainable solutions. Part I explores the overwhelming evidence for the reality, promise and ongoing suppression of free energy research, based on two decades of first-hand experience and documentation of over a century of proofs-of-concept. In Part II, I weave the environmental mandate we all face into the coming energy revolution, which I then argue is the only viable answer to humanity's current global predicament. In Part III, I confront the roots and rationale behind the tyranny whose purpose is to move us further away from the world of abundance which free energy could provide. I also call for unifying our divided and ruled cultures through our combined positive intention. *The Energy Solution Revolution* is a wake-up-call to transform planetary culture from one dominated by capitalistic self-interest, pointless wars, ecocide and dirty energy, into one where the bulk of humanity can at last come together with compassion for all creation—before it's too late.

This war perpetrated by elite corporations and the U.S. and other rich oligarchies against humanity and nature has only escalated over the past few years. The unfolding of these crimes provides a stark contrast to how things could be in a better world. We need nothing less than profound systemic change in our political, economic and social systems in order to have clean energy, and that must happen sooner than is comfortable for most of us. But our discomfort and disease coming from an imperiled biosphere would be far worse. We need a revolution, a peaceful transformation. Its centerpiece should be a clean energy solution revolution that could restore sustainability and sensibility to the world.

During the 1990s, I visited the laboratories of over a dozen inventors worldwide of electromagnetic devices that appeared to produce more energy than went into them, as measured by traditional means. What we are seeing is energy coming from the vacuum of space, sometimes called zero-point energy, because it still exists at temperatures of absolute zero. I've seen many of these proof-of-concept demonstrations for myself. I wrote up the results of this journey in my 1996 book *Miracle in the Void*. Since then, many more books, scientific papers, articles, organizations and websites describe the prospects of new energy technologies that could reverse a permanent state of war and the gloomy spectre of irreversible climate change, the drastic pollution of our air, water and land, and the seizure of dwindling nonrenewable resources such as oil, coal, natural gas, uranium, forests, water, food, soil, minerals, wetlands, coral reefs, pristine diverse natural habitats and indigenous cultures.

Many of us with a scientific background have also investigated revolutionary concepts such as cold fusion with palladium cathodes (nonradioactive, room temperature nuclear reactions), sonoluminescence (acoustic cavitation, a cold fusion method), and special hydrogen and water chemistries, all of which are almost certainly producing significant clean excess thermal energy in the presence of catalysts. We have come to the conclusion, based on sound experimental science, that we could soon have a breakthrough energy economy, if we only allow these courageous researchers to move the technologies beyond their own laboratories. Unfortunately, that has not happened yet. The global controllers' suppression of making available any practical device has been 100% airtight, in spite of numerous attempts to break through. Covert assassinations, threats, thefts, draconian laws prohibiting the patenting of such devices and lack of funding have all thwarted development.

This book does not explore the details of the concepts themselves, specific suppression stories nor the mechanics of how to enact widespread public demonstration and development. Numerous writings in the scientific literature and on the Internet address these issues in depth, and can be accessed through some of the websites listed herein or through Google. Many technological assessments, startup companies and inventions flood the Internet, and for those of you who want to learn more about the technical credibility of these ideas, I suggest you look carefully at the unfolding evidence for their efficacy. I can assure you that most of this unfoldment is not disinformation!

If you're still skeptical or don't want to look at the evidence, then I ask that you suspend disbelief and merely explore the possibility that our future energy could be abundant, clean, cheap and decentralized. In such a case, how, then, should we implement and regulate these new technologies? That is what we will be looking at in this book.

A few more disclaimers: I've noticed in my numerous interactions with the public, that what many people expect of me, I no longer do. Some people expect me to tell them whether or not I've seen a UFO, whether or not we landed on the Moon, or if we already have our own free energy device in our home. Some want me to show them the principles and mechanics of the most viable new energy concepts, to provide them with a complete, quantitative airtight explanation and proof of this or that invention and its theory, to get in on the ground floor of the enterprise, and to give them the best considered information about where they should put their investments into, and when they can trot down to K-Mart to be the first on their block to buy their own free energy gizmo. Some individuals want me to "show them the beef" before exploring the possibilities any further.

I am sorry to disappoint those of you looking exclusively for solid facts, proofs or disproofs. I'm out of the vetting business, in part because many other technically trained and up-to-date researchers whom I've gotten to know and trust are doing that job very well. Also, I really don't know which particular technology will take hold or when, in today's complex economic and political environment. In addition, the process of scientific research, which is also a process of scientific "search," can yield many different possible pathways to success. Edison tried thousands of times to perfect the light bulb, even though he had proven its usefulness several trials earlier. So it is with solution energy.

This trial-and-error process of convergence is not as important as the fact that many energy solution technologies do exist, any one or some of which could provide by far the best answer—if the climate of suppression can end.

As we explore the issues in this book, we shall see the problems and solutions to our energy problems are not technical. They are human. The evidence for new, breakthrough, "over-unity" energy is overwhelming, but we must realize that we are in the research phase of a research and development cycle, that we are on the toe of the profit curve that almost no venture capitalist yet wants to touch, and that so far, the inventors have been divided and conquered by the unmitigated greed of the existing energy lobby.

Our challenges are therefore social, economic and political. We will see that, like thrashing dragons wanting to get one more drop of blood, we crave hydrocarbon and nuclear fuel in deference to those who wish to profit from our continuing use of dirty energy and who are doing everything they can to stop solution energy from seeing the light of day. We shall see that the pervasive use of carbon trading, biofuels, alternative hydrocarbons such as the tar sands of Alberta and oil shale of Utah, carbon sequestration at coal mines, the hydrogen economy, nuclear power, hybrid cars, air cars—even solar and wind—can only distract us from what we really need to do in the long run. We shall see that these "solutions" are just smoke and mirrors in the well-publicized energy sideshow. As journalist George Monbiot put it, the first way to keep from environmental and climate disaster would be to keep carbon (and uranium) in the ground. But then what?

Policymakers and an ignorant public are thwarting authentically clean energy at every turn, either by commission or by omission. As a sociological problem, the need for truly clean energy is perhaps the most urgent one our culture has ever faced. We'll be looking into these dynamics of overcoming our denials and vested interests as we progress through this book. Given that both policymakers and the public commons have difficulty supporting solar and wind power, multiply that suppressive force by millions when it comes to the viability of new energy technologies to save the planet from almost sure disaster.

We will explore the fact that almost every "expert" who presents the "renewable" alternatives limits his/her discussion to the known sources such as solar and wind, which are not really all that renewable. The public is deceived into thinking those are all there are or ever will be. The free energy possibilities are simply discarded at our risk and peril.

My journalist-colleague Keith Lampe has well expressed a scenario for our future energy choices. He suggests that existing "renewable" technologies such as solar and wind be implemented as first-generation energy, while we develop forthwith second-generation energy such as zero-point, cold fusion and advanced hydrogen technologies. Both options, he feels, need equal time, whether it's in Congressional hearings or in public discourse. Looked at in this light, we can evolve an energy policy that makes sense, and could give us a much better chance to reverse climate change before it's too late.

I agree with Lampe's assessment, which was also the principal conclusion of my own studies, as expressed in my previous book *Re-Inheriting the Earth* (2003). It seems that, even for those skeptical about whether or not these advanced energy technologies are real, is it not wise to adopt the precautionary principle that we leave no stone unturned to uncover those energy sources which could really solve our paralysis of paradigm and save humanity and nature from almost certain destruction? We cannot afford to do less.

But we must also ensure that the technologies are regulated by responsible government and are never put to weapons use or overuse. That's why social change must go along with introducing any radical new technology such as this one.

A word about terminology: I have used many adjectives interchangeably in describing what I mean by revolutionary new energy technologies: free, new, clean, solution, breakthrough, advanced, over-unity, innovative, novel, unconventional, revolutionary, outside-the-box, second-generation, zero-point, vacuum, cold fusion, novel hydrogen, etc. For the purposes of this book, they all basically mean the same.

They each represent a potential quantum leap in our ability to have clean and cheap energy. In every case, using these sources can give all of humanity an authentically lasting new paradigm in our energy policies and practices, at first involving disruptive and seemingly magical concepts. They give us an opportunity to transcend humanity's rampage on nature and give us a reasonable chance to have a truly peaceful, just and sustainable future on Earth—if we can take action soon and wisely.

This is the kind of book that's best to read in small bits. I recommend your reading one or two chapters at a time, and let at least one night pass between them. This approach would allow you to have more time to gestate, to ponder the enormous implications about what we can do if we only look. I'd give this reading two weeks minimum at several minutes per day to an hour or two. After reading and revising the text, I felt this book is a bit like a course, one I hope will lead to new ideas and action.

Some of you might want to know about the state of the art of free energy for presentation to general audiences. For that, I recommend Chapter 12 "Open Letter to Al Gore," Chapter 13, "A Word to Innovators about an Energy Solution Revolution" (this chapter was earlier published as an article on the U.K. World Innovation Foundation site, of which I was recently elected a Fellow), Epilogue and the Appendix, which is a peer-reviewed paper I presented on a wide range of alternative energy options at a conference in South Korea in 2007. These four sections are what I would call the primers for understanding the possibilities of practical free energy in a digestible way for the political mainstream.

I have used the word "political" in the subtitle of this book because politics at its best should be about translating what's scientifically and technologically real, and potentially beneficial, into the realm of possibility and action. But in these times, politicians in the U.S. government are enormously corrupt and are oblivious to sustainable solutions. They will need to become responsible to the people instead of to the elite corporations and financiers who support their campaigns. We can research and vet all the energy options we want and we can even try to do this job honestly, but none of this will matter unless we can effect political action at many levels. We're going to need to know the truth and find ways of acting on it constructively. We must search for a public context to introduce solution energy for all.

The best way to do that is to begin a frank, open public discussion of how we can best bring forward these clean energy technologies and do an end run around the likes of the warmongers and arrogant capitalists in charge. We don't want Dick Cheney to be running this one too. In my opinion, the prospects of free energy depend entirely on the success of an altruistic democracy that is longing to be born.

– Brian O'Leary
Vilcabamba, Ecuador
October 6, 2008

Part I

Breakthroughs and Suppressions

Prologue

Confessions of a Naïve Scientific Heretic:
A Story about the Carrot and the Stick

"In a time of universal deceit, telling the truth becomes a revolutionary act."

– George Orwell

"When stupidity is considered patriotism, it is unsafe to be intelligent."

– Isaac Asimov

I have a close friend whom I'll call Ted, who has taken a trip to hell and back. After many years of suffering, he's now sane and happy again in his personal life. Yet he's still grieving about the imminent demise of civilization. After I heard his story of intrigue, pain and almost certain death, I myself began to tremble in fear. This was too close to home, I thought.

You see, Ted is a scientific heretic. So am I. He had been a straight-arrow with a distinguished background in government service and academic research. So was I. He is an idealistic visionary civilian working on controversial alternative research on healing ourselves and the planet. So am I. He was also naïve and sometimes gullible. So was I. And we both have suffered a lot, whether it be from threats, bad health, bankruptcy, ridicule, isolation and loss of career security.

I guess you could say he came back into my life at the right time. It gave me a chance to reflect on the journey of a heretic, to try to get a grip on why we as a culture would rather perish in scarcity, poverty and violence than go for lasting solutions.

I began to explore the principle that, sooner or later, heretics might confront powerful people, especially if the heretics are effective in their work. At the very moment of Ted's going public with his outside-the-box ideas, some high-up government agents dangled a carrot in front of him. They wanted to recruit him for a lucrative secret project in his area of expertise.

Unbeknownst to Ted at that time, the code language of the "offer" really meant that this was an assignment he couldn't or shouldn't refuse. He didn't know about that. Yet it didn't feel right to accept even though he found it a bit tempting to join the team for the sake of the country and to reap some financial rewards as well.

Following his gut and his heart, Ted answered "no" to the proposition. A short time after, his life was to descend into a hell of hurts and personal traumas, not the least of all was an attempt on his life. He lived in pain from the episode and in fear of what might happen next. But Ted's return to the living-normal gave him a sense of gratitude that he could still make a difference in the world. Ted was lucky and also the wiser about looking for signs of any new carrots and sticks coming his way. Most of all, he felt free and happy that he turned down the offer.

Ted's story opened me up to contemplate all the other war stories of my courageous colleagues, many of them free energy inventors and spokespeople who had been assassinated, bullied or bought out.

And as Ted's story unfolded I got in touch with a deep sense of grief about an unacknowledged and life-threatening force so unimaginably evil that no individual or part of nature was safe. The unwritten message to Ted was, "If you live, civilization and nature will die. But if you become too effective in saving civilization and nature, then *you* die." A Faustian Bargain for the innovators, at best.

My contemplation led to what might be one of the most radical and yet believable (to me) conspiracy theories of all: if we do our healing work well, someone will either point a gun to our heads (and maybe shoot it) or give us a bribe to *keep quiet*, to cease doing our work if we want to stay alive. Sometimes they can even order us to help them do their dirty work.

The logic of this book clearly points to such an extreme possibility. The facts I present, alongside Ted's story, painfully revealing as they may be, forced me out of my own naivety and into understanding some basic principles of revolutionary work on free energy and other perceived threats to the status quo:

1. Most scientific heretics can be offered bribes, threatened, attacked or assassinated almost anywhere, any time. There are hundreds of ways to murder someone and make it appear as a suicide, accident, natural death or isolated assault. We are all very vulnerable.

2. Elements of the ruling class often commit these acts when they feel their control is being challenged. The homicides go unpunished almost every time.

3. The perpetrators take their orders from a shadowy cabal that is organized like the Mafia. Their motive is the consolidation of power feeding off vested interests in controlling the world's resources, energy, military, drug trade, big agribusiness, economy and banking infrastructure.

4. Distracted by bread and circuses, most people are ignorant or fearful of these truths and thus deny or do not support the possibility of breakthrough solutions such as free energy. "Most people," writes Shaun Saunders, "prefer to cling to the belief that the 'state' will only do what is best for everyone—a childlike frame of mind, that encourages turning a blind eye to many evils...to think outside of this comforting box is too frightening for them, and would require that they take responsibility for what happens in their lives and the events that transpire around them." In this regard, democracy is dead.

5. Only by strength-in-numbers will we be able to overcome this tyranny and build a world of peace and abundance. That means we shouldn't expect to wait for mainstream scientists or commercial interests to bless and deliver finished products to us.

We return to these principles in Part III in looking for ways to overcome the darkness. Meanwhile I'll describe the breakthroughs, their suppression, the scientific mandate to change our ways, and the fact that free energy may be the *only* option available to us for a sustainable future for civilization.

Chapter 1
Pigs Can Fly!

"If the Scientific Establishment trusts only in its textbook theories and if they disbelieve people of good will who have the means of bringing (new energy) forward and choose 'not to look through the telescope,' the consequences will be that these wondrous technologies will not be developed as rapidly as they would otherwise—or they may not be developed at all! This has been and will be a monumental tragedy for virtually every category of human experience, all of which will be transformed by these now apparently 'unwanted' discoveries."

> – Eugene Mallove, www.infinite-energy.com, 2004

"The resistance to a new idea increases as the square of its importance."

> – Bertrand Russell

Over one hundred years ago, two obscure bicycle mechanics from Dayton, Ohio, flew the first airplane. One would think this event and many to follow would have been reported and celebrated around the world.

They weren't. One journalist from the Wright brothers' hometown newspaper was fired from his position for reporting such a heresy. More than one year later, *Scientific American* ran an editorial debunking the whole idea because newspapers didn't report their repeated flights. We had a chicken-and-the-egg problem: Scientists didn't believe anything that wasn't reported, but it wasn't reported because no mainstream editor or scientist would accept it.

Shades of Galileo's colleagues refusing to look through his telescope: This is the conundrum of scientific and engineering discovery: new developments cannot "exist" unless "credibly" reported. It took five years for the Wrights' achievements to become 'credible' in spite of thousands of eyewitnesses to several flights. The spell of denial lifted when Theodore Roosevelt ordered official test flights in 1908.

Our time is no different, but the stakes are higher. The issue now is clean energy and planetary survival. Hundreds of independent experiments on a wide range of new energy concepts have been successful but remain unreported and unsupported in the mass culture beyond the Internet (as we shall see, many sites give a very different story from the newspapers and electronic media). Any one of these could change the world. Given the proper public support, renewable, clean, cheap, compact and abundant power packs and gas cells could soon become available. We'll need no more hydrocarbon fuels and nuclear power plants. We'll need no more human-caused climate change, no more air pollution, no more water shortages because of cheap desalination of oceans, no more starvation, no more poverty, no more resource wars, no more fascism.

I grew up with a can-do attitude in the spirit and tradition of the Wright brothers. As a schoolchild during the late 1940s and 1950s, I had aspired to go to the Moon and planets when there was no space program; in 1967, I was appointed by NASA to be the first scientist-astronaut to go to Mars. I had been an Eagle Scout, climbed the Matterhorn, taught astronomy and physics at Ivy League universities, and led marches on Washington to help end the War in Vietnam.

The genocidal debacle of Vietnam was new to me. It ended the Mars program and served as a warning to all Americans of what was to come. Times began to change. Sociologists have pegged the 1950s as a time of greatest happiness, confidence, prosperity, optimism and security in the United States. Unknown to me then, it was also a time of "the birth of unbridled consumerism, and in my mind, the beginning of a most dangerous and possibly fatal era: the birth of a society that would eventually be best described by 'groups of one,'" as noted by Shaun Saunders. Previously, we had suffered two world wars punctuated by a depression. And, over the next fifty years, we have suffered a steady moral decline of staggering impact, committing the war crimes of genocide and ecocide with the utmost greed and empire, epitomized by the Bush administration. In a few short decades, we Americans got to be cast as the bad guys, and for good reason.

So my formative years coincided with some of the best years in U.S. history, and they modeled for me what we could do, that we could and would solve any problem that came our way. I know this at a very deep level, probably as most younger people don't know now, and are understandably cynical about a system deeply ridden with violence and hubris.

Through the years, my optimism had been battered so much, I found myself shifting jobs and careers, hoping to make a difference in the face of disappointment, human folly and corruption. It became difficult for me not to be pessimistic about our future. I still have a spark of hope, however, that innovations in energy could help save the day. But the stage is set for a monumental David-and-Goliath, death-and-rebirth kind of struggle, to bring these and other outside-the-box solutions to a bewildered and frightened culture in denial.

My pessimism is well-founded, because the prospect for an energy solution revolution has been suppressed at every turn by powerful vested interests. The media again passes while mainstream scientists wallow in denial for fear of ridicule ("if it isn't reported or properly vetted by vested money and intellectual interests, it isn't real"). The result is an unwitting alliance between establishment scientists and the corrupting energy barons and their governmental and media mouthpieces. Meanwhile, we continue to be addicted to oil, so much so we don't seem to know a good thing when it comes along. Yet, most of us know, at least at some level, that we need to transform this addiction to chain-smoking our oil and coal and move on to alternatives before it's too late. We must lift the contradictory veil of credibility.

From ten years' direct experience at witnessing new energy breakthroughs in laboratories around the world, I can personally vouch for the successes in solution energy research, whether it be cold fusion, advanced hydrogen chemistry or vacuum energy. But, like during the Wrights' first flights, we are not delivering the product yet. We are in the research phase of a research and development cycle. The research, if properly supported, will inevitably lead to the deployment of energy systems that will profoundly change the world.

Why can't we perceive the truth hidden beneath the conundrum of credibility? It seems that credibility is simply a fantasy created by media, academe, politicians and corporate interests. In this game of smoke and mirrors, style has usurped substance, moreso than ever in these trying times. Hidden under the radar of the mass culture, we are missing out on concrete solutions, with the truth lying not so far below, but actively suppressed by current powers, who see such developments either as impossible or as a threat to an economy based mostly on polluting, destabilizing and unsustainable energy resources. Politicians rarely see beyond the next elections and corporations rarely see beyond their next earnings report. Both powers use immoral accounting practices to line their pockets.

I am convinced we could have a comprehensive energy policy leading to near-zero emissions by 2020. The research is mature enough to set this goal, just as JFK had done for the Apollo lunar missions. I am also convinced that a publicly funded R&D effort of some hundreds of millions of dollars will catapult us into a sustainable future with many energy choices. On the other hand, we can maintain our cultural "credibility" by doing nothing.

Meanwhile, the research goes on in scattered locations by inventors in government labs, universities or on their own, with little or no support or acknowledgement from the government or the scientific mainstream. In my opinion, the development phase needs to become transparent and public. It is too important to be left to existing powers whose economic self-interest is suppressing solution energy at every turn. Yet we may need it to avert global disaster from pollution, climate change, prolonged blackouts, wars over oil and fiscal thievery.

We can learn another lesson from the development of aviation. Even though the Wrights and many others were repeatedly flying during the early 1900s, support for the technology awaited World War I in 1914. The same happened with the deployment of nuclear energy during World War II: military applications drove the technology.

Will we have to wait for a military application of new energy before we see it? We cannot let that take place. This technology is too important to be left to the military-industrial-secrecy complex. I already shudder about what would happen if we supported the U.S. military's insane notion of "Full Spectrum Dominance" by deploying weapons in space while blanketing the Earth in ever-more lethal forms. Likewise, I hope to never see the day when military elements control new energy technologies.

Some of us are seeking ways of educating the public about breakthrough clean energy research and its application under the guidance of democratic systems. We need to facilitate free and open discussions about the best choices. We must minimize price-gouging and empire-building and military secrecy. Let's begin a wave of personal and environmental freedom triggered by a new clean, cheap and decentralized energy system.

The Wright brothers have already flown in the new energy game. We now await delivering passengers, mail and cargo, and we cannot expect the inventors to do that. Nor should we defer to large energy companies and the existing government to make wise choices. The public must take control of this issue and guide us through the transition to a Clean Energy Age. Only then can we educate and empower ourselves to come up with the best ways of creating a bright future for generations to come.

Between 1990 and until his untimely murder in 2004, the late Eugene Mallove, former chief science writer at MIT, editor of *Infinite Energy* magazine (www.infinite.energy.com), and tireless crusader for new energy, had often reported on the degree of ignorance, arrogance and fraud among mainstream scientists on the veracity of cold fusion. His incisive critiques came from the startling discoveries of University of Utah chemists Martin Fleischmann and Stanley Pons and the many peer-reviewed replications and conferences. His magazine, thousands of peer-reviewed scientific articles and annual meetings involving dozens of academic and governmental scientists from all around the globe have attested to the reality of cold fusion.

Soon after the release of the Pons-Fleischmann discovery, Mallove began to uncover fraud in the negative results nuclear physicists at MIT reported in trying to replicate the original work. They furthermore labeled the apparent breakthrough as a "scam" and "scientific schlock". These new critics were not even trained in electrochemistry as were Pons and Fleischmann, whose techniques were totally different from those of the nuclear physicists.

The deeper answer to these dilemmas comes from the vested interests of the nuclear physicists, who have received tens of billions of dollars to attempt to control "hot" fusion reactions within gigantic devices called Tokomaks. After more than forty years of trying to confine a plasma gas within magnetic fields, they have been unable to create an energy "breakeven". With generous funding but no results, these people rose to the top of establishment science, nevertheless. One of the nuclear physicist debunkers was Charles Vest, who later became the president of MIT. Talk about vested interests...

The next step in debunking cold fusion was obvious. The U.S. Department of Energy (DOE) hastily convened a panel made up mostly of hot fusion physicists who appeared to nail the lid of the coffin of cold fusion. According to Mallove, this report had the following consequences: No special funding by the U.S. government for further research, flat denial by the U.S. Patent Office of applications mentioning cold fusion, suppression of research on the phenomenon in government laboratories, and citation of cold fusion as "pathological science" or "fraud" in numerous books and articles critical of cold fusion in general, and of Fleischmann and Pons in particular.

It's ironic that many of the leading cold fusion scientists continue to carry on their work quietly at such disparate locations as the U.S. Los Alamos Labs, the Naval Research Labs, the China Lake Naval Weapons Labs, Oak Ridge Labs, the University of Illinois, the Japanese Tsukuba Space facility, Hokkaido University, the French Atomic Energy agency, and numerous others. In spite of impressions to the contrary, cold fusion continues to be a robust science.

Nevertheless, one of the original DOE panelists, Professor and Provost Steven Koonin of Caltech, took exception to the exception to a rule, "his rule." Like the editorial pundit for *Scientific American* one hundred years ago, Koonin stayed inside his box so much, he can only hurt the cause of an energy solution revolution.

Koonin declared, "My conclusion is that the experiments are just wrong and that we are suffering from the incompetence and delusion of Doctors Pons and Fleischmann...it's all very well to theorize how cold fusion in a palladium cathode might take place...one could also theorize how pigs would behave if they had wings. But pigs don't have wings."

Ergo, pigs can't fly, right?

Wrong. What Koonin neglected to say is that pigs don't need wings, but they could become passengers in airplanes, albeit probably not in first class.

Ergo, pigs can fly!

Some of the material of this chapter is taken from *Re-Inheriting the Earth* (2003) and an essay "New Energy: on the Centennial of Aviation: How "Credibility" Overwhelms Truth".(2003).

Chapter 2
Who's Doing the Suppressing?

"Since the time Ronald Reagan was elected President in 1980, the global energy-ecology challenge has been clearly swept under the rug and it is now largely perceived in political circles as a nonissue. This is very sad…"

– Miracle in the Void, 1996

"The people of the world need to know that the oil companies have become the enemies of mankind. They are withholding and suppressing technologies which could save the planet and eliminate poverty and suffering. As long as we allow there to be a "secret government" this will continue to be so. Those who toil in basements and garages to bring forth energy from the vacuum in spite of death threats and assassinations are the true heroes of this age."

– Thomas Bearden, 2007, www.cheniere.org

"The House of Representatives is one big steaming dungheap that should be leveled and turned into an amusement park instead of a tax-payer-funded knocking shop. What a pathetic collection of cowards and scumbags."

– Mike Whitney, www.informationclearinghouse.info,
Sept. 21, 2008

"There is an irreconcilable conflict between the goal of creating economically just and environmentally sustainable societies and embracing sustained economic growth, unregulated markets, and free trade as the organizing principles of public policy. The resulting policies are well suited to producing more millionaires and billionaires… These are real world consequences of an out of control financial system in which reckless young traders backed by the massive financial assets of leading private financial institutions send billions of dollars sloshing around the world in a high stakes gambling frenzy with an almost complete absence of oversight."

– David C. Korten,
When Corporations Rule the World, 2nd Edition, 2008

In 1993 and 1994, some of us founded the Institute for New Energy, which held two "think-tank" retreats and public symposia in Estes Park, Colorado. Researchers attending both meetings presented ample evidence for the theoretical and experimental efficacy of free energy, which has the potential to replace traditional energy in the near future. At that point in time nobody had succeeded in making these devices available for individual use, and the blackout for development continues to this day. We were, and still are, in the research phase of a research and development effort.

The governments and private industries of India and Japan at that time had funded top-level scientists to develop breakthrough energy for commercial applications, something about which the American government appeared to know little or nothing about outside of the black budget. Things now have gone even more underground, as the distracted U.S. Empire collapses and seems to be pulling the rest of the world down with it. The oligarchs in charge are creating an illusory scarcity consciousness and gobbling up the planet's precious resources as if there were no tomorrow.

For awhile, it looked like the development of new energy would happen off-shore. Cold fusion pioneers Martin Fleischmann and Stanley Pons, formerly of the University of Utah, moved to France to be funded for their work by a Japanese consortium. The inventor of the N-machine, the late Bruce DePalma, formerly of MIT, further developed his free energy concepts in Australia and New Zealand. Other American inventors and researchers have gone underground most of the time (e.g., Thomas Bearden and the late Sparky Sweet), have been sued (Sweet), had their devices taken away by impatient investors with gold fever (DePalma), had their machines confiscated by the government (e.g., the Canadian inventor John Hutchison and American Dennis Lee), have been convicted and jailed under questionable charges (Lee) and in at least one case have been told by the Government to change careers—or else (e.g., Adam Trombly).

More serious have been the ongoing suspicious deaths and

injuries suffered by solution energy inventors. Eugene Mallove was brutally murdered in an apparent robbery. University of Graz physics professor and leading European free energy scientist Stefan Marinov "jumped" from the library building to his death. Water-fuel car inventor Stanley Myers collapsed and died after running out of a restaurant shouting, "They poisoned me!" Journalist Gary Vesperman has compiled 53 cases of the deaths and serious injuries and illnesses suffered by new energy researchers (www.rense.com/general72/oinvent.html).

As I write this in November 2007, Arie DeGeus, inventor of a self-powered battery, with considerable near-term promise as a compact solid state free energy source, was recently found slumped over and unresponsive in his car at the long term parking lot near the Charlotte, North Carolina, airport. He died shortly thereafter of a heart attack. Probably no coincidence, he was on his way to Europe to get support to develop and commercialize his invention.

All these men and many others were in the prime of their lives. Cutting-edge new energy researcher Tom Bearden postulates that DeGeus was murdered by an electromagnetic gun called a "Venus shooter." This device destroys the body's control of its heartbeat. Bearden himself and several others have been victims of these assaults, some of which were fatal, and some not. The common denominator to these heart attack victims was that they also could be attributed to "natural causes," thus concealing the crime.

Whether by "suicide," "accident," "natural causes" or "homicide by a lunatic," this carnage of our foremost unsung heroes seems to have a common cause: assassination by those who don't want breakthrough energy (or any other expression perceived as a significant threat to elite vested interests) see the light of day.

I am sickened and grieve deeply each time a leading revolutionary thinker with a positive vision is taken out. These tragic deaths join the ranks of others, including those of John, Robert and John Jr. Kennedy, Martin Luther King, Malcolm X, Paul Wellstone, Mel Carnahan, Salvador Allende, David Kelly, and countless others in politics, science and popular movements, pioneers who could have made a real difference.

To ponder the violent genocide of our heroes and innocents reveals the most grievous crimes of our time, crimes that go unpunished by international and constitutional law, either perpetrated or ignored by a corrupt system that is rotten to the core. What has happened to our morality? Why have we allowed the foxes to guard the chicken coop? Why have they gone to such extremes? Why are we afraid of confronting these criminals? We return to this most important question in Part III of this book.

In all, I have met several dozen free-energy researchers. What all these individuals have in common is the underfunding of their work such that it proceeds to proof-of-concept but no further. Developing useful prototypes requires a much larger effort. This could come from bringing the researchers together in a research and development team, analogous to the Apollo or Manhattan projects. But there has been no public and little private support for free energy inventors—particularly in the United States—even though this country is where most of the ideas come from. We seem to be so active in suppressing this technology, that we have driven most of our brightest inventors away or underground.

Many of us in the new energy field came to the 1994 Colorado think-tank filled with the kind of optimism I had felt during my youth. While Jim Carrey was filming *Dumb and Dumber* at our hotel, some of the best and brightest breakthrough energy inventors flew in from all over the world to meet and strategize about a well-funded future for everyone involved. Researchers Thomas Bearden, Moray King, Paramahamsa Tewari, Shiuji Inomata, Harold Puthoff, John Hutchison, George Hathaway, Toby Grotz, Patrick Bailey, Maurice Albertson, Robert Siblerud, Don Watson, Wingate Lambertson, Harold Aspden, Peter Graneau, Stefan Marinov, Andrew Michrowski, Hal Fox, Tim Binder, Dale Pond, Don Kelly, Jeane Manning, John Searle and Ken McNeil were among those attending.

Our optimism was well-founded because the software hundred-millionaire who flew everybody in was about to announce the foundation of a new corporation that would support many of the most promising concepts for commercial development. When the sponsor of the conference approached the podium near the end of the meetings, we were eager to hear what he was going to say. But, as we shall see in Chapter 6, he delivered another stillbirth, this time triggered by the reluctance of the venture capitalists to enter the fray before they could climb the steepest slopes of the profit curve (snort).

The airtight denial and non-support apply as much in today's world as they have been for decades, and the suppression goes on— with some exceptions during the proof-of-concept stage of development. Some of the more promising concepts and their state of the art can be found on the website www.newenergycongress.org, where several experts vote on the top one hundred technologies being researched.

The remarkable fact is that we seem to have already demonstrated this technology for more than one century! Nikola Tesla, followed by hundreds more energy mavericks, have repeatedly demonstrated free energy, only to be suppressed before the technologies could be developed for practical use. For one hundred years, we probably didn't have to pollute the Earth to meet our energy needs!

In my studies of the suppression syndrome of our times, most all members of the following communities are keeping us away from the most important new developments in science and technology:

The scientists themselves. Throughout history the curious phenomenon of scientific denial is what prevented Galileo's colleagues from looking through his telescope, or what caused the French Academy of Sciences to deny the existence of meteorites, what has made contemporary astronomers ignore the UFO evidence, or what made Harvard University want to appoint a secret committee to fire the late John Mack or for Princeton University to want to fire Robert Jahn, in both cases, scientists willing to explore topics taboo to the academic establishment. In his classic text *The Structure of Scientific Revolutions*, Thomas Kuhn has clearly shown that scientists can be very unscientific when their own world views are threatened by colleagues.

Industrial suppression. Oil and utility executives generally do not want free energy to happen. General Motors did not want electric cars and innovative batteries, and so they buried the patent. Pharmaceutical companies and doctors do not want to see their profits end with a miracle cure. Banks and insurance companies don't want to handle any big change. In each case, a multibillion dollar vested interest perceives a threat to its own well-being. We return to this topic in Chapter 5.

U.S. Governmental secrecy. This alphabet soup of agencies (CIA, NSA, DIA, ONR, etc.,), born and bred from World War II and superpower confrontations, spends untold trillions of the American taxpayers' dollars to sustain an unaccountable, suppressive and aggressive force that is now abusing its power. The crimes of the executive branch, as epitomized by the Bush II administration, are but the tip of an iceberg of an enormous black budget nightmare of genocide, torture, domestic spying, fear-mongering and fascist control. Because it is sometimes impossible to distinguish information from disinformation, and because secrets are compartmentalized and based on need-to-know, those who have investigated this hydra-headed beast believe that the Cosmic Watergate of UFO, alien, mind-control, genetic engineering, weather engineering, eugenics, secret prisons, mercenary armies, tasers, weapons in space and super-weapons on Earth, antigravity propulsion, 9/11 truth and other secrets will make Watergate or Irangate appear to be kindergarten exercises. Unfortunately and unforgivably, the U.S. Congress, judiciary and mainstream media remain complicit, while the rest of the world looks on with horror and disgust, but which most Americans forget quickly.

The media. In suppressing the suppression stories (or relegating them to the Internet or tabloids, lumping these all into one category they call "conspiracy theories") the media have abrogated their responsibility to report the truth. Most all of them are a meek, cynical and self-aggrandizing lot, operating under strict guidelines about what to investigate and what not to investigate. They are stenographers of the lies and edicts of their corporate masters. Free energy, miracle cures and UFOs are intrinsically big news, but ignored by those charged with informing the public because of a perception of lost credibility, the safeness and party atmosphere of pack journalism, and the element of control and censorship by publishers, networks, the government and others holding the strings of money and power.

Ourselves. This is probably the largest single factor. At some level subconsciously, we do not want free energy. We're afraid. "Better the devil we know than the devil we don't know." Sad to say, you'd think that those of us with an environmental and progressive outlook would be open to second-generation energy. But most of us fall back upon our own traditional renewable energy biases toward solar and wind power. These people seem to want to deny the credibility and feasibility of big breakthroughs, out of fears of losing their own hard-earned and poorly-supported turf, and of the potential for weapons use or overuse. They also seem to be overcome with a sense of scarcity consciousness and a lack of vision. Yet these environmentalists and progressives, if properly educated, could become allies. In future chapters, we shall see how that might be possible.

As recently as fifteen years ago, I was in denial about free energy. Twenty-five years ago, I denied UFOs. My bias came because I erroneously assumed I would have known about it if it were real and that my scientific colleagues know best. The only way I could begin to lift my own veil of denial was to see it for myself, to meet the people making the breakthroughs, to gain a scientist's credibility check of both theory and experiment. Only my own self-realization about the potential of these extraordinary possibilities brought me beyond the denial. Only our collective realization about these developments can bring our culture out of its denial. As in the "hundredth monkey" fable, when enough of us have gone through that process, we will all have a revolution in consciousness. We need a critical mass first, however.

Having written close to one hundred cautious peer-reviewed papers in the scientific literature that have withstood the test of time, my acceptance of the radical contents of my newer research did not come to me casually. I felt that, in order to understand reality, I had to travel the world many times to visit and see for myself the experiments and demonstrations performed by those talented inventors, scientists and psychics who are the true pioneers of our time. Along with them, I have often sensed the alienating effects of ridicule, economic deprivation, lack of identity, anger and depression.

Many inventors such as Bruce DePalma, John Hutchison, Sparky Sweet and others have shared with me their own sense of isolation. We are not alone in weathering this process; it is part of being a human on the front lines of change.

But it is worth it! By its very nature, free energy is democratic, a symbol for getting us out of our imprisonment of paradigm, which clearly is not where we need to go if we have a decent chance to survive and thrive.

In the next few chapters, we shall see how most scientists, environmentalists and industrialists seem to want no part in acknowledging, let alone supporting, the potential of an energy solution revolution. This gives us a chance to contemplate entirely new strategies and tactics for getting beyond our long-standing social gridlock.

Some of this chapter includes updated excerpts from my book *Miracle in the Void.*

Chapter 3
Eugene Mallove Versus the Stuckness of Scientists

"No problem can be solved from the same consciousness that created it."

– Albert Einstein

"I have had fifty years of experience in nuclear physics and I know what's possible and what's not...I will not look at any more (cold fusion) evidence! It's all junk!"

– Herman Feshbach, MIT professor, 1991

"I hope you recognize that the late Professor Feshbach's most unfortunate and ill-considered reaction was fundamentally unscientific. It reminds me of the Church leaders at the time of Galileo, who refused to look through Galileo's telescope at the Moon or at Jupiter because they "knew" that nothing new could be seen. Yes, many modern scientists are filled with catastrophic hubris; they have in many ways become mere "technicians of science," and guardians to what amounts to a pernicious "Holy Writ." Don't bother me with the experimental evidence, my theory can tell me what is possible and what is not!"

– Eugene Mallove, www.infinite-energy.com, 2004

"For the first time, experimental evidence was presented that shows conclusively that excess heat can be reproduced on demand...on May 22, 2008, Professor Yoshaki Arata at Osaka University (presented) experiments in which the ratio of output to input heat has exceeded 2500%...

– Scott Chubb, *Infinite Energy*, September-October, 2008

M ore than any other source of breakthrough energy suppression, some of our most powerful scientists lead the way. The world looks up to these high priests of our culture for their yea-or-nay pronouncements of whether or not any innovative energy is possible. Their answer is almost always "no" without their even giving it a look. We are caught in the conundrum of credibility. Only a few of us with a scientific background have broken free of this kind of censorship, and can recognize the great promise of new energy research.

Nobody knew and expressed this potential better than Dr. Eugene Mallove—scientist, journalist, educator, visionary, humanitarian, friend. On May 14, 2004, he was murdered during an apparent robbery in the home of his parents. He had been in the prime of his career at age 56.

This is a great loss, which is just beginning to sink in. So prolific were his writings, so passionate was his expression, so unique were his connections between the scientific community and the solution energy vision, he will be sorely missed. He is one of my heroes. It will be difficult for anyone to fill his shoes.

As chief science writer for MIT in 1989, Mallove first got into the new energy game by investigating the controversial discovery of cold fusion by University of Utah chemists Drs. Martin Fleischmann and Stanley Pons. As we saw in Chapter 1, his investigative reports on the attempts of MIT nuclear physicists with a vested interest in hot fusion research (consuming tens of billions of federal dollars) revealed that they fraudulently reported their inability to replicate the original Pons-Fleischmann experiments. Their "unadjusted" findings actually reinforce the truth of the discovery. The efforts of these scientists and other establishment researchers at major universities to discredit cold fusion, under the auspices of the U.S. Department of Energy, will be remembered as a sad lesson in scientific ethics.

Risking everything, Gene Mallove resigned his position at MIT to reveal the truth about cold fusion regardless of ridicule and financial insecurity. He wrote the book *Fire from Ice*, which was nominated for a Pulitzer Prize. The facts revealed in that work led the great writer Arthur C. Clarke to conclude that the suppression of cold fusion was "one of the greatest scandals in the history of science."

Mallove had been at it for fifteen productive years, before he was senselessly cut down, what a loss! His authoritative books, his numerous editorials and writings in the successful magazine Infinite Energy, his support of the work through the offices of his New Energy Foundation and New Energy Research Laboratory, his many media appearances, and the excellent video/DVD "Fire from Water" underline the boldness with which he struck out on his mission, fearless about threats to his career which otherwise would have been secure. He stood alone in organizing and supporting the cold fusion scientists, which has grown into a community of several hundred researchers, many of whom came from prestigious institutions, meeting annually and publishing in the best peer-reviewed scientific journals. Against all odds, cold fusion has become a viable science with great potential because of the tireless and selfless efforts of this one man. He was a giant.

To underline my remarks, just visit the web site www.infinite-energy.com. In my 2003 book *Re-Inheriting the Earth*, published before his death, the index shows more pages of citation to Mallove's work (nine) than that of any other individual. He was very grounded and did not venture into consciousness science as I had done. This made him all the more credible. He never succumbed to the pseudo-scientific claims of new age pundits and acted as a responsible scientist throughout. He didn't play favorites but looked at all viable alternatives with intellectual honesty, yet still from a Western scientific point of view.

His standards were always very high. He did not buy into fads and politics. His rigor could be very disarming to those toeing the line of the establishment and was an inspiration to all fellow heretics. I not only stand in awe of his achievements, I miss the opportunity to team with him and others to raise the credibility of new energy research and development.

He had also proposed to the Democratic presidential candidates in 2004 the establishment of an Office for Unconventional Energy within the Department of Energy (DOE). During a candidates' panel he orally presented the idea and three of the candidates endorsed it. We can largely thank Gene that the DOE began reviewing the cold fusion concept, something that would have been unthinkable during the past fifteen years.

His "Universal Appeal for New Energy" is posted on Google and on many websites. Please read it. Its articulate expression is unmatched in our quest to give this research a chance. He has been the standard-setter of our work, providing a foundation for understanding the scientific principles and nomenclature for the most promising concepts and categories.

His grasp of historical analogues such as the development of aviation, relativity, quantum physics, the Copernican revolution and the

Enlightenment, as reported in *Infinite Energy*, are inspiring and pro-lific. His writings provide a deep and insightful understanding of the historical patterns of scientific progress and censorship that we now face in new energy research. I humbly defer to Mallove's writings and to the brilliant classic book by Thomas Kuhn, *The Structure of Scientific Revolutions*, about how mainstream scientists in any culture can be blind to new paradigms of discovery.

When it comes to changing worldviews, the scientists, alongside vested government, industry, media and intellectual discourse can become the guardians to the gates of outmoded thinking. More than anyone, Mallove gives this cultural aberration, applied to new energy breakthroughs, fresh insight that will endure for a very long time, par-ticularly as the developments much too slowly unfold—sometimes in direct contradiction even to the thinking of progressives and environ-mentalists, who are imprisoned in scarcity consciousness.

The tragedy of Eugene Mallove all the more inspires us to advance our mutual cause. We must get the word out as quickly and widely as possible. Most of us in the field believe that this murder was an assassination: he was one of many of those pioneers who could make the greatest difference. This mandates that, through strength in quality and numbers, the teams now forming will ensure that the work will carry on regardless of the personal fates of pioneers such as Gene. A full investigation of the murder, still pending, will hopefully shed light on its true motive.

Novelist Mignon McCarthy said, "Each society honors its live conformists and dead troublemakers." I believe that the world will remember Dr. Mallove as one of the greatest scientific pioneers of his-tory, a Galileo of our time.

In later chapters, we'll look more deeply into how to overcome the stubbornness of those who call themselves "scientists" who refuse to look at new phenomena that appear to violate their own precepts of reality. The scientists, as we shall see, can be allies or adversaries in our quest. For example, the climate scientists and leading environ-mentalists have helped us create a challenging new mandate to keep carbon in the ground. This is good. But most physicists still resist the notion of free energy, even though the evidence is overwhelming. They too are of the no-free-lunch school. We shall have to bring more and more scientists, environmentalists and progressives on-board.

The material of this chapter was based on the eulogy of Eugene Mallove, "Galileo of our Time" (2004), and published in *Infinite Energy* Magazine of that year.

Chapter 4
Clean Breakthrough Energy and the Environmentalists...
Is There a Meeting Ground?

"Ladies and gentlemen, I have the answer! Incredible as it might seem, I have stumbled across the single technology which will save us from runaway climate change! From the goodness of my heart, I offer it to you for free. No patents, no small print, no hidden clauses. Already this technology, a radical new kind of carbon capture and storage, is causing a stir among scientists. It is cheap, it is efficient and it can be deployed straight away. It is called....leaving fossil fuels (hydrocarbons) in the ground.

– George Monbiot,
The Guardian/U.K.,
Dec. 11, 2007

C ontemplating the staggering transition from the oil economy to a solution energy economy invites disbelief, however essential this change must be for our own survival. Most mainstream environmentalists are among those avoiding these issues. They seem to be too busy putting out fires and wallowing inside the box of scarcity-thinking, a critique the late Buckminster Fuller often expressed as a devastating lack of vision that could do us all in. Or could it also be, at a subconscious level, the ecologists couldn't trust the implementers not to mess this one up too? Instead of a tiger in our tank, do we have one by the tail? Do we have to assume we cannot take control of our own collective destiny?

Many greens have avoided new energy like the plague. No wonder, given our energy policies, derided as "bungled" over the last three decades even by corporate media giant *TIME* magazine. Our energy policy is the culmination of accelerating industrial and political vested interests that are surely leading us all to extinction either by pollution or by war—or both.

The ecologists are right: how could anyone trust the energy establishment to manage new energy? How could we simply stand by while corporate profit-centers decide for us whether to go with coal, oil, gas, tar sands, oil shale, nuclear power, solar, wind, biofuels, hydrogen, fuel cells, etc. The bottom line is that breakthrough energy does not appeal to big business because of its simplicity, cheapness, renewability, and decentralized nature.

I wholeheartedly agree with the progressive greens' critique, and support strong action against our current practices, for example Monbiot's brilliant suggestion to leave the oil, coal, etc. in the ground. Yes! But he only solves half the problem—letting go of what we don't want. But what do we want in its place? What's next? Overcoming the lack of awareness of the possibility of clean and abundant solutions, especially among those who correctly see a crisis in search of answers, is a thread that will run throughout this book.

Common sense dictates a strong disgust with the American government's mandate not to reduce its emissions just to feed the appetite of the energy and war industry's thirst for ever-more business, power and influence. Most environmentalists and I can also demonstrate that many government-and-industry-imposed "cleaner" alternatives such as carbon sequestration at coal plants, spewing particles into the atmosphere, and burning biofuels just to marginally mitigate emissions so we can keep gassing up the billion private vehicles, are very bad ideas. Eating and injecting bad stuff into the biosphere and burning our food, are the crazy notions worthy of a society-gone-mad searching for band-aids. So, too, is a hydrogen fuel cell economy, which can only serve an elite few because of its great expense and its own energy requirements to produce the hydrogen in the first place.

Most of the educated world understands that these excesses can only exacerbate rather than reduce global pollution. But is the misappropriation of effort skewed towards the needs of the corporate culture any reason to block promising new possibilities for the rest of us? Why must we accept conventional wisdoms about our energy future, based on the self-interest of industry and government, and at great cost to the rest of us?

The news coming from the December 2007 UN Climate Change Conference in Bali is not good. The U.S. and China lead the way away from agreeing to the modest Kyoto emissions reductions, while promoting carbon trading as the only solution to our dilemma. As we shall see, these kinds of neoliberal gimmicks give the biggest polluters the right to burn more, not less, in this world of flush windfall profiteering.

We have three communities objecting to breakthrough energy: (1) the powers-that-be, the current energy industry and its cronies in government, combined with a secrecy apparatus that covers up the truth of new energy, while the Pentagon spends bigger budgets and produces more weapons than ever; (2) mainstream scientists and their students, whose interests are best served by defending the familiar old turfs of thermodynamics, nuclear physics, and the denial of the existence of energy from the vacuum of space or from novel catalytic reactions with hydrogen; and (3) most environmentalists who are scared of the potential misuse of new energy and are steeped in the zero-sum game of scarcity, and so deny it on bogus scientific grounds, or they hope the problem will go away. Most of the rest of us remain in ignorance.

As one green economist put it, in a new energy future, we would have our skies swarming with personal helicopters like locusts, bigger bulldozers, weed-whackers and power saws...and even more awesome weapons of mass destruction. We don't want that!

So this begs the question, how can free energy be regulated? Which applications are benign and which have the potential to do great harm to nature or be abused as weapons or overused by overconsumption? No matter which choices we make for the future, we need to be selective about which ones make the most environmental sense rather than be victims of the winds of corporate and governmental power. We need to consider a wide range of renewable options, ones we will look at in Part II.

Where I differ from some of these environmentalists is that we can proceed towards a solution energy age by: (1) carefully selecting those sources which can deliver energy on small scales with no weapons potential, and (2) reasserting public control of our energy choices. Like ordering from a menu at a restaurant, we can select whatever energy system we'd like—in principle.

We may need to limit the amount of wattage available to customers, commensurate with current uses at first, and with later adjustments according to need and to the full life-cycle environmental impact of the energy sources used—that is, the cost of extracting, transporting, manufacturing, purchasing, using and disposing of the devices. A more elegant approach, suggests Saunders, could be: "The other side of the solution is to move away from the competitive, ego & self-esteem driven consumer paradigm which calls for bigger plasma screens and the latest and biggest of everything. That is, remove the demand aspect from the supply and demand equation." Of course, the problems of reversing our economic tyranny are central to the needed transitions to free energy.

But the dialogue has just begun, and some of the traditional environmentalists will need to broaden their education about both the potential uses and abuses of breakthrough energy. Among the spokespeople and organizations who will need to expand their list of choices beyond incremental first-generation solutions are Al Gore, George Monbiot, Ross Gelbspan, James Howard Kunstler, Robert Redford, Richard Branson, Richard Heinberg, Julian Darley, Amory Lovins, Paul Hawken, James Lovelock, Michael Ruppert, Kelpie Wilson, Harvey Wasserman, Hazel Henderson, Bill Mckibben, John Holdren, Steven Lewis, Michael Klare, Ralph Nader, Critical Mass, the Union of Concerned Scientists, the Natural Resources Defense Council, Sierra Club, Friends of the Earth, Greenpeace, Climate Crisis Coalition, the No War No Warming alliance and many others.

We hear from these individuals, groups and their followers a lot of the stern warnings and wake-up calls about the crisis we are in. But we get little about long-term solutions, and nothing about the new energy technologies, except for the occasional dismissal that it can't exist because the scientists said so. I call this the conspiracy of the SEPs, the paradigm paralysis of institutional scientists, environmentalists and progressives. Author of the brilliant website www.ahealedplanet.net, Wade Frazier, recently expressed that their rationalist-materialist philosophy, scientism and "conspiracy-phobia" make these individuals a tough nut to crack. They can only join ranks of the other guardians to the gates of change and so will need to go through their own paradigm shifts. I'm not holding my breath about that kind of breakthrough without further crises and education.

The situation is even worse than that. The U.S. environmental movement itself has lost most of its political effectiveness in the past decades. Since 1980, when Ronald Reagan was elected U.S. president, the greens have gotten weaker while the polluters have gotten stronger. Decisions are more often settled by compromises in boardrooms rather than confrontations in courtrooms. Money and career advancement become a greater motivation for influencing policy than doing what needs to be done. This watering-down and taking baby steps couldn't even come close to solving the physical realities of global pollution and climate change. Author Mark Dowie well documents this loss of a robust green movement in the book *Losing Ground* (MIT Press, 1996):

"Unlike the other social movements of the 1960s and 1970 (women's, peace, civil rights, and gay liberation), which are essentially radical, the ecology movement was saddled with conservative traditions, by a bipartisan, mostly white, male leadership... rarely have they challenged the fundamental canons of western civilization or the economic orthodoxy of welfare capitalism. The ecologically destructive system gives the nation's (added comment: now, the world's) resources away to any corporation with the desire and technology to develop them."

The co-option of environmental movement by "conservation corporatists" actually began from much earlier, around the time of Earth Day in 1970. According to Keith Lampe, one of the authentic founders of a truly radical environmental movement, "The U.S. environmental movement began with several humans risking their lives for trees and has degenerated into a non-movement of comfortably salaried ecocrats. The corporate wing has dominated the movement since the mid-1970s. During this time the condition of the biosphere not only has worsened—but has done so at ever-increasing rates."

The escalating crisis calls for taking much more intelligent and decisive action. As we shall see in Part II, we have no other option but radical action. Unless the greens can move beyond their warnings, lamentations, and harangues about imminent and inevitable disaster and into solutions, they will have to pass on the leadership to new blood, a whole new movement, regardless of whether or not we adopt breakthrough energy technologies.

The world simply cannot allow human excesses to spill over to a breakthrough energy domain. In my opinion, we are going to have to be smart about how we apply free energy and we are going to have to regulate it, like any other energy source. Or would you prefer that an Enron or ExxonMobil monopolize and manipulate energy markets over what you had thought to be solution energy? Of course not, unless you happen to be one of the immoral few who can profit from the energy oligarchy now controlling things.

Is fear alone going to prevent us from doing what will be necessary to stop us from our crazy dependence on oil, coal, natural gas and nuclear power? Most traditional renewables such as solar and wind are better, but they're old and incremental. They can be expensive, intermittent and materials-and-energy-intensive. In our search for cheap, clean and benign energy, we must leave no stone unturned.

Most of all, we can trust neither the existing energy industry nor their governmental cronies to control all this. Their current dominance is killing us all, and so they must be deposed and their technologies phased out. Nor can we allow them exclusive rights of ownership of novel forms of energy. For our own survival and that of all living systems, we can thank the old systems for having delivered energy and then let them go—entirely. Neither corporate welfare nor the denial of the most promising alternatives will do. We will have to change.

To these ends, researchers at www.newenergycongress.org are compiling a database of second-generation solution energy concepts, with applications that could be clean, cheap, renewable, decentralized and benign. These technologies will comprise a menu from which the public will be able to order, under responsible regulation already discussed. We believe our search will uncover several hundred—perhaps several thousand—previously suppressed technologies which will prove to be eco-friendly in all respects.

In later chapters we explore what we mean by benign as a criterion for all energy applications. We must become free of the control by those seeking power and profit. We must also become free of the scarcity consciousness which can only feed the greed of those who go to war for dwindling resources while impotent environmentalists holler "foul" but go no further because their own belief systems are the same as the plutocrats: there isn't enough to go around.

Yet I still hold out hope this kind of new energy policy will win over the greens of all stripes and we will be well on our way to a zero emission society much sooner than any conventional thinking would suggest.

As we shall see, the urgency of the issue demands a new kind of activism that transcends the traditional conservation movements, who are now losing ground. Our very being depends on this kind of paradigm shift. In the next chapter, we look at the most significant obstacle of all: the endless search for corporate profit and quest for resources.

Portions of this chapter were adapted from the essay New Energy and the Environmentalists and the book *Re-Inheriting the Earth*, both published in 2003.

Chapter 5
New Energy in the "Free Market" Age:
Will large corporations help or hurt us?

"Corporations have been enthroned...an era of corruption in high places will follow and the money power will endeavor to prolong its reign by working on the prejudices of the people...until wealth is aggregated in a few hands...and the Republic is destroyed."

– Abraham Lincoln

"(When a working class youth seems to be) the intellectual equal of the rulers, a difficult situation will arise, requiring serious considera- tion. If the youth is content to abandon his previous associates and to throw his lot whole-heartedly with the rulers, he may, after suitable tests, be promoted, but if he shows any regrettable solidarity with his previous associates, the rulers will reluctantly conclude that there is nothing to be done with him except to send him to the lethal chamber before his ill-disciplined intelligence has had time to spread revolt...Every (governing class) youth will be subjected to a threefold training: in intelligence, in self-command, and in command over oth- ers. If he should fail in any one of these three, he will suffer the ter- rible penalty of degradation to the ranks of common workers, and will be condemned for the rest of his life..."

– Bertrand Russell, from *Scientific Technique and Education,*
pp. 244 and 248
and quoted in www.knowledgedrivenrevolution.com.

"We've been suckered again by the U.S. So far the Bali deal is worse than Kyoto. America will keep on wrecking climate talks as long as those with vested interests in oil and gas fund its political system...The whole U.S. political system is in hock to people who put

their profits ahead of the biosphere…Until the American people confront their political funding system, their politicians will keep speaking from the pocket, not the gut."
                    – George Monbiot, *The Guardian/UK*, 17 Dec. 2007

I read with interest an op-ed article "The Courage to Develop Clean
Energy" by Jeffrey Immelt and Jonathan Lash in the 21 May 2005
Washington Post. The title seemed to come right out of the theme of
the first conference of the New Energy Movement in September 2004
in Portland, Oregon entitled "New Energy: The Courage to Change."
Without yet identifying the authors of such a seemingly prophetic
piece, I read on.

"After inventing the light bulb," the essay begins, "Thomas
Edison was asked where he grew inspiration from. 'I find out what the
world needs,' Edison replied, 'then I proceed to invent.'" Who are
these enlightened beings, I asked, that are able to speak our language
and be able to break through into the mainstream media as well,
denied to the rest of us in the innovative energy movement for such a
long time?

Eagerly, I read more. Immelt and Lash proposed that three ingredients were necessary to deploy clean energy: "(1) the brainpower to
develop new technology, (2) a market that makes clean technologies
profitable, and (3) a strong dose of American will." Immelt and Lash
then argued we had the first two ingredients in place, but we needed
to develop the third.

I disagree. The requirement for profit, under some conditions,
could actually eliminate many promising options which are either not
ready for the market yet, or are intrinsically cheap. Hence we have
Corporate America's emphasis on more mundane and massively sub-
sidized and deployed technologies such as nuclear power and ethanol-
from-corn, which could quickly turn in the profit. Hmmm, who were
these columnists anyway?

My eyes then flashed down to the writers' biographies. Jeffrey
Immelt is chairman and CEO of General Electric Company and
Jonathan Lash is president of the World Resources Institute. Heavy
hitters…

What can we learn from their statement? In my opinion, we all
have a lot in common, except for the profit "requirement." What
kinds of profits? Who decides how much profit is enough? To satisfy shareholders, any company like GE must turn in a humongous
profit in the tens of billions from the dirty energy and war machinery

fields to grow and thrive. All this assumes, then, that any kind of energy innovation must come under a capitalistic umbrella controlled by large corporations. The implication here is that, if a clean energy technology were basically free, then it is not worth GE's or any big company's while to press forward with the technology. They would pass and, if necessary, suppress. If cheap energy were to ever threaten their bottom line, this becomes central to the conspiracy to bury promising new technologies (and sometimes bury the inventors themselves!).

So let's ask, what if a given commodity which was highly polluting but profitable (e.g., oil, coal, gas, biofuels and nuclear power) could be replaced by a very cheap clean technology that would turn in very low profits? Would GE be happy if it had to give up its nuclear and gas turbine power plants for this? Of course not. The Immelt-Lash article clearly implied that no large corporation vested in the energy sector and accountable to its shareholders could give up its profits for something less profitable. In fact, the authors glossed over the rationale for their profit requirement and made the quest for growth axiomatically true for all energy technologies at all times. They then shifted the blame for our poor energy track record to what they see as the lack of the third ingredient: a strong dose of American will. I believe, to the contrary, that the first and third ingredients are there but latent, and that the capitalistic axiom is what is blocking us and promoting inadequate substitutes that would hardly make a dent in slowing the inertia towards a planetary catastrophe.

Up to this point, we have looked at objections to a new energy future posed by mainstream scientists and environmentalists. But this one recent article coming from the bowels of the U.S. corporate establishment, while giving lip service to innovation, can effectively veto concepts that don't turn in sufficient profits. Breakthrough energy would be a prime candidate for a veto. Meanwhile, the CEO class can blithely blame the American people for their lack of will. This is very tricky and dishonest.

Some corporations have become so powerful they have set our energy policies purely out of the profit motive and have joined at the hip with the U.S. government through massive campaign contributions to both sides of the aisle. Since 1990, the dirty energy-resource exploitation sector has given over $400 million to Republicans and Democrats. During the 2000 campaign, while Bush got the lion's share, Gore still received almost $500 thousand from these companies. We might expect that politicians like Gore to arrange for a company like BP to erect expensive solar panels, with government funding and kickbacks, of course. Rampant corruption, power-brokering and fighting resource wars are the inevitable outcomes.

But this issue profoundly affects the health of the global commons. We cannot allow wars and privatized resource grabs and greedy energy use to dominate our decisions any more. When it comes to energy, war and water, the public will need to learn how to take its power back, to awaken so we can steer the ship of state away from a titanic catastrophe. That process is now well underway here in Latin America and needs to re-enter the consciousness of the American people at this critical juncture.

Of course, in the real world, the "profit axiom" does dominate and is the greatest source of the suppression of new energy. I'm not suggesting we shouldn't make modest profits from new energy developments. But profits cannot be the pacing item in bringing in an energy solution revolution. Alas, many of us in the U.S., particularly, look at the world through the fuzzy filter of privatization as a panacea to world economic challenges. Can we trust the testimonies and rationales of those now in power? Or do we truly have the courage to address fallacious assumptions about policies that would require turning in huge profits?

The corporatization of basic human needs has made a mockery of Adam Smith's original thinking about free markets on small scales among equals. Instead it has led to the dangerous accumulation of private power warned about by so many former U.S. presidents including Jefferson, Lincoln, both Roosevelts and Eisenhower. The CEO of General Electric can wax eloquently about clean energy innovation and blame the American people for their lack of will, but does GE have the courage to tame (I'm not saying eliminate) the profit motive in the event of significant new energy breakthroughs?

I have an open question for Mr. Immelt. Hypothetically, what if GE were to be invited to manufacture 10 billion clean 10-kilowatt new energy power packs to be sold to the whole world for $10 apiece, turning in only a small profit, certainly not on the scale of GE's conventional power plant systems? Would GE do it or pass or maybe suppress the new technologies? What other corporations would want to be involved and how? Is this not a question of cooperating to save the planet instead of its opposite: business as usual among giants hooked on growth? Could not GE go to the government and get assistance in transforming their priorities to clean ones, leaving growth out of the equation? Herein lies the greatest dilemma of unmitigated capitalism. Maybe it was no accident that, in drafting this chapter, I had misspelled Mr. Immelt's name as Mr. Innert.

Answering this question could give us a key to our collective survival. And if Mr. Immelt or any other CEO's answer perchance were to be yes, then modest profits truly could combine with ethics and I'd like to collaborate. I'd also like access back to the media that has effectively blacked out any talk about radical energy solutions. But this hope becomes unlikely if the GEs and dominant governments of the world aren't willing to reject their old priorities and convert them to the new.

Meanwhile, we can be very wary of making much progress with vested big business, and I know much of this would seem heretical to the American way. But we need to create a new context for change that transcends the optimization of profits. For example, I helped presidential candidates George McGovern in 1968 and Jesse Jackson in 1988 propose massive programs to convert the aerospace war machine priorities into ones that could produce clean energy and other peaceful projects that would employ people on government contracts. Of course, neither of the candidates nor their ideas succeeded, because the profits for such endeavors would be less than building weapons and fighting wars.

The need for a radical shift in our war and energy practices is more urgent than ever. Not doing so will surely lead us into disaster—one that is obvious to more and more of us. Taking the discussion to its logical conclusion means that General Electric, Boeing, LockheedMartin, ExxonMobil, ChevronTexaco and the other Pentagon and energy giants, alongside the entire captive U.S. and multinational financial, governmental and media infrastructure, would kill (and do kill) to suppress clean energy so these entities can not only survive, but thrive solely for the benefit of its management, investors and employees.

After all, that's exactly what this plutocracy is now doing on a massive scale, by fighting aggressive wars for oil and destroying our freedoms and ecosystems. In a word, their rampage is literally the crime of murder. And the controllers would prefer us all to go down with the planet to get their own share. To them I urge: think again. Get out of this addiction, go somewhere else, don't destroy the biosphere as you're doing, don't erect smokescreens. Come clean. An energy solution revolution requires a systemic revolution on a global scale that is so massive our dominant institutions must cease to exist in their present form for the rest of us to have even a prayer.

Meanwhile, the world certainly cannot wait around for a General Electric or a government to solve our problems for us. They got us into this mess in the first place. They will have to sacrifice their vested interests for the greater good, or get out of business.

We have a few remedial options, all of which will involve the collapse of the current economic system (which is beginning to happen anyway) and the concomitant emergence of clean energy development under any one or more of the following models:

Altruistic funding of new energy R&D by a sympathetic government or private benefactor(s). This model could be open (like the Apollo Moon project) or secret (like the Manhattan project). Whichever way this goes, the project must be for the benefit of the people, no weapons application. For years, I had thought that one or the other would have logically moved forward, but now that the entire system is so corrupted, I am not sure such a research effort could begin without the already great security risks toward those involved. Funding for developing the dozen or two most promising technologies would be on the order of $1 billion, less than one day of Pentagon spending. The U.S. Department of Energy is so corrupted that there is not the slightest interest in funding the work. Nor is Congress. Richard Branson, are you listening?

Establish a mutual fund for new energy R&D. Shortly before his untimely death, Eugene Mallove invited the investment community and general public to pool their money in a long-term, high-risk, high-gain fund that could immediately support the work of the leading researchers. Altruism could combine with the possibility of getting on the ground floor of the planet's most important paradigm shift. Again, this would be a logical choice in a more rational world, but also has its own security risks.

Open-source the core technologies. This proposal forwarded by Joel Garbon, president of the New Energy Movement, is modeled on the successes of the computer industry. The standard business model includes invention and prototyping, patents, raising investment capital, selective licensing with royalty provisions, manufacturing at centralized locations, and established marketing and distribution routes. The open-source concept short-circuits this chain, whose weakest links have historically stopped development. Instead, at the point of invention of the core technology, no patents are needed, it's basically given away so that the inventors and others may benefit from developing applications for commercial use. Technologies would be demonstrated at high profile press conferences in key cities globally.

There'd be simultaneous wide-broadcast dissemination of "how-to-build" plans on the Internet, and the public would be encouraged to use and improve the technology and create useful applications (no licensing or royalties). This approach is akin to the altruistic one, which sidesteps the almost-impossible task of patenting a break-through energy technology and the politics of corporate burial and governmental and academic ridicule. The public exposure is very important, of course, to bring the core concepts into the realm of transparency missing in our current cultural gridlock. This would allow the technologies to come forward as quickly as possible.

Go off-shore and develop solution energy under sympathetic new democracies unbeholden to empires and oligarchies. Many governments in the new century, startled by the tyrannies of our times, have taken a more independent path and are creating innovative social systems to encourage their people to support free, open, healthy and sustainable practices. Some governments with a long history of violent repression have emerged with enlightened new leadership. For example, the recently elected governments of Rwanda and Ecuador (where I live) would encourage rather than discourage a solution energy revolution. Other emerging economies such as those of China and India also have every reason to move in this direction. For example, Dr. Paramahamsa Tewari, a former Indian government physicist and chief engineer of the Kaiga nuclear power plant in Karwar, India, was given time and laboratory space to test his new space power generator, based on the principles of Bruce DePalma's N-machine.

I have only one caveat for proceeding with any of the above: we will need to regulate the technologies so their use would have minimal environmental impact and no weapons application. We don't want to repeat the mistakes of nuclear energy. This becomes an important challenge for the legal and parliamentary communities of all nations: to ensure this technology is forever for the benefit of mankind and the Earth. No more war toys or excuses to over-pollute or overuse. We return to this vexing question in Chapter 17.

We have seen that an innovative energy economy could transform our whole energy paradigm. A billion small power packs that produce clean and renewable electricity and as many magic cells that can heat water and air, upon further development, testing and evaluation, could be history's most important single technology for our benefit and that of nature. This issue is too important to those who require any form of massive profit, as dictated by a General Electric. In cooperation with a visionary leadership and public, they are going to have to let go of all of that, surrender their thirst for the machinery of war and dirty energy. If they want to survive as a group, they will have to join the crowd that longs for a future of authentic sustainability and abundance. Some things need to be free and energy is one of them.

Portions of this chapter were adapted from the 2005 essay of the title as this chapter.

Chapter 6
Dumb, Dumber and Dumbest:
Reviewing the Myths of
Breakthrough Energy

"I recently had a conversation with a learned man who recited the many potential disasters that humanity is poised on the brink of, largely due to our energy practices. I agree with that assessment. When I mentioned that free energy could eliminate all those problems, and quickly, he dismissed it in an instant, saying that we could not predict how free energy's implementation might develop. He preferred certain doom rather than potential salvation, because he could not predict the final outcome, even though the upside was something that looked like heaven on earth. I have witnessed that kind of response dozens and dozens of times."

– Wade Frazier
www.ahealedplanet.net/hooked.htm

"We have an ongoing program for the development of technologies which utilize esoteric energy for propulsion including the zero point dynamic of vacuum space and various multidimensional theoretical developments offering sources of new energies previously unknown."
– U.S. Department of Defense:
Program Solicitation for FY 1986, p. 193,
AF 86-77 and subsection 6 (unclassified)

In May 1994, about fifty new energy researchers gathered at the historic Stanley Hotel in Estes Park, Colorado in the hope of forming a team to pursue our common interests in breakthrough energy development. We had hoped that this think tank would give us a chance to get smart about a sensible solution energy policy, to found a "skunkworks" to bring energy research into the 21st Century. Yet we have little to show for it ever since.

We had expected a lot from the software almost-billionaire who flew us in from all over the world, the second such conference he had convened in two years. He had promised support for the best and brightest of inventors, leading to commercial new energy. But when his marketing people admonished him from funding an effort "not yet dipping into the river of optimized profits," he backed out. We were still in the toe of the profit curve, they reasoned. "We don't want to throw good money after dead-ends. Let's wait until we have a winning horse." And I later learned this was the law of venture capitalism. It was strategically too early to invest in breakthrough energy. The under-funded researchers went home disappointed and empty-handed. Of course, it was also possible that the powers-that-be placed effective carrots and/or sticks in front of our would-be funder to withdraw his support.

The photo of our group included an anomaly: Jim Carrey. While we were trying to be smart and smarter, Carrey was filming *Dumb and Dumber*. Perhaps this improbable confluence of minds meant something beyond our own intention to organize ourselves. In the fourteen years since this historic meeting, we have made very little progress in supporting the suppressed research or in educating the mainstream culture towards the enormous potential of solution energy. We are still curiously dumbed-down in a conundrum of censorship from which we have still not emerged. And we continue to worship some myths that rival those in Orwell's *1984* or the twisted pronouncements of the Bush administration. Here are some of the most outrageous myths:

Myth #1. New energy is not scientifically valid.

This is perhaps the biggest obstacle of all. Throughout history, mainstream scientists have opposed and ridiculed innovation. They are the guardians to the gates of truth. The more important the idea, the greater is the resistance. Solution energy research is no exception.

I have talked with fellow physicists still in the mainstream who are open enough to discuss the matter with me. To them, it would be professional suicide to advocate breakthrough energy research because that implies the acceptance of revolutionary concepts frowned upon by colleagues. "Any claims that appear to violate established theory or any attempt to rigorously address new principles would embarrass me. My funding would stop. It's not worth the risk for me to stick my neck out until I'm already 100% convinced the experiments would work." This is a classic chicken-and-the-egg conundrum.

Eugene Mallove expressed the situation this way: "If by chance you are one of those who believe that 'all is well in the house of science' and that 'official science' can be counted on to behave itself and always seek the truth—even in matters of central, overarching importance to the well-being of humankind—you are sorely mistaken, and I could prove that to you with compendious documentation."

And so, the typical physicist waits silently and skeptically until somebody somewhere might produce a convincing (to them) experiment that they hope won't work. Then a slow change can begin, but not one moment sooner. Meanwhile, one government-funded physicist, Harold Puthoff, has been sitting in his laboratory awaiting the arrival of research devices, one at a time, for testing. Is this any way to run a crash program for planetary survival? Certainly not: the results so far have been (understandably) disappointing and slow when looked at by institutional physicists.

Regardless of such fallacious perceptions, several successful experiments in low energy nuclear reactions (aka, cold fusion) have been reported in the peer-reviewed scientific literature, for example, methods of producing energy through acoustic cavitation or sonoluminescence, by multiple authors from prestigious institutions worldwide reported in *Science* Magazine in March 2002. Many other approaches appear in the literature as well, in spite of the denials of some mainstream physicists, and in spite of little or no funding available to do the work.

Some physicists simply hide behind the "laws" of thermodynamics and deny the existence of "perpetual motion" as their rationale to naysay new energy at the outset. What they don't seem to understand is that these theories or concepts apply only to closed systems in equilibrium and for which energy conservation is understood in a limited context. Ilya Prigogine won a Nobel Prize for pointing out these limitations in his chaos theory. Sad to say, it is difficult to convince many embedded physics professors about these violations. It's like talking with them about the paradoxes of quantum theory and nonlocality: they know these things are true, but they don't want to talk about it. As a result, the U.S. Patent Office still denies applications that relate to free energy.

A prime example of how the old science keeps gripping us in its reactionary clutches was when I testified in July 2003 before the

California Energy Commission hearings on reducing California petroleum dependence. While most witnesses proposed incremental solutions reflecting their own commercial interests, I spoke about the evidence for a coming energy solution revolution with free energy technologies. In an ironic twist of the occasion, the chairman of the Commission was Arthur Rosenfeld, one of my old U.C. Berkeley graduate school physics professors during the 1960s. After the hearings, Professor Rosenfeld, a specialist in thermodynamics now in his eighties, came up to me and said that the best we could *ever* do in improving energy systems was to make minute improvements in the efficiencies of converting existing engines. There was no free lunch, he declared, there was no possibility of energy from the vacuum, no chance of catalytic reactions to trigger novel chemical or nuclear processes. End of discussion.

But, as Gene Mallove once wrote: "'The exception proves the rule.' Or put another way: 'The exception proves that the rule is wrong.' That is the principle of science. If there is an exception to any rule, and if it can be proved by observation, the rule is wrong." It is sadly poignant that these and Mallove's other statements quoted in this book were some months before he died, in a prophetic document entitled, "Open Letter to the World" describing his "Universal Appeal for Support for New Energy Science and Technology." (www.infinite-energy.com, 2004).

In spite of all the positive evidence for innovative energy, some of you might still believe Mallove and I are totally wrong about this, that we will never have practical free energy because "there is no such thing as perpetual motion." O.K., I grant you may never change your mind and will go to your grave with and bequeath your beliefs—just like London's still-functioning Flat Earth Society. To you, I say the same as I say to the climate crisis skeptics: inaction would have far more severe consequences for the biosphere if you and the other skeptics are wrong, than the investments required for taking action if we new energy advocates were wrong. Logic dictates that the "precautionary principle" trumps the endless debate about whether or not climate change or breakthrough energy is real. Based on overwhelming evidence, it's time to move on and act on what is really happening to us and what we can do about it.

Myth #2. Extraordinary claims require extraordinary evidence.

This scientific double standard is a corollary to the first myth. Popularized by the late Carl Sagan, this is the credo of skeptics enforcing scientific orthodoxy in this era of the (feared) deconstruction of mainstream physics. It is simply a defense mechanism to deter new ideas, in which the Occam's Razor goalpost is arbitrarily and politically moved ever more towards the skeptical view: "In the absence of countervailing evidence, the simplest explanation shall suffice." The countervailing evidence is more often ignored by many physicists, setting up lack of support for the research.

Such reasoning flies in the face of the search for the truth, sliding the scale of credibility to suit the political and economic agendas of those in charge. It is absurd to demand more experimental evidence according to whether the question to be asked happens to be important. Mr. Sagan was wrong. We must go where the evidence leads us regardless of our biases as to how ordinary or extraordinary the question might be.

Myth #3. If it were real, we'd have it by now.

This is the classic conundrum of invention, recalling the early days of aviation, when Scientific American pontificated that heavier-than-air flight couldn't be real because otherwise it would have been reported. But it wasn't reported because editors didn't believe it could be real, even though there had been thousands of eyewitness sightings. The myth basically posits, "Because it isn't on the news it cannot be real." But the news itself is censored. You can begin to see the contradictions. This is sheer dumbness.

A corollary myth is that the research is so exotic, any commercial application must be further off in time, like a science fiction scenario for the 22nd Century and beyond: therefore, it's best to continue doing what we're doing, to wring out all the oil and build massive conventional energy infrastructures, even "renewable" ones such as wind, solar, the hydrogen economy and the use of biofuels, all of which have their own challenges. But focused research in innovative directions could actually bring on the new in shorter times than any vested interest would want us to know about.

Myth #4. We must await the magic bullet.

Some of us who tend to be complacent have grown to believe that, some time in the distant future, one savior-inventor might come forward to rescue the world with his device. Until then, it's business as usual, and even then, anything new will have to wait its turn. This myth basically says, "We may some day have a Bill Gates to come forth competitively. The magic of the free market will make all this happen in its own wise timing, and may the best man win. If or when all this happens, I want to be the first on my block to have it." This simplistic belief also ignores the decades of suppression of promising directions by those in charge.

This all-or-nothing thinking plays right into the hands of powerful corporations who can pick and choose energy systems that best fit their own profit projections. This has nothing to do with planning sensible energy policies that are in the public interest. This is just plain dumb.

Myth #5. We can trust the government to support clean energy R&D.

This might be true in a more rational world. But the current leadership of Western governments has, in concert with powerful industrial and media lobbies, blocked development of these technologies, except within black budgets. Even solar and wind power have not received the public attention they deserve. Is this any way to run an energy policy? Certainly not: the public will need to reassert its will.

The material of this chapter is based on an essay first posted in 2004.

Chapter 7
Solution Energy Truth and Lies of Omission

"Historically, efforts by 'new energy' pioneers have been met with 'resistance' (to put it mildly) from global oligopolistic energy (mainly oil) interests. Furthermore, many new energy pioneers and alternative energy advocates have tended to focus narrowly on pet theories or projects to the exclusion of valid sources promoting a comprehensive and real solution to the energy problem. Indeed, these efforts have combined to make obvious truths, alternative viewpoints and possible solutions unthinkable. The reality is that new energy exists, is unlimited, forever renewable, non-polluting and can free the world from a paradigm of scarcity and introduce it to one of abundance."

– Wade Frazier,
www.ahealedplanet.net/scarcity.htm, 2004

George Orwell has said that the biggest lies are lies of omission. The manifold manipulations of the Bush administration and its media mouthpieces brought this deception to a fine art. They skillfully managed our perceptions by selecting themselves as the sole "newsmakers" on important issues while avoiding any deeper context or greater truth.

Hidden from view are both drastic corruption and exciting possibilities that lie outside any serious consideration by the collective mainstream. But the consensus perception is just now coming to grips about the problems, which will inevitably lead to understanding their deeper roots. This is a process of "truth and reconciliation" which is only beginning to reach the awareness of some people and select politicians.

We can begin to sense the principle that "if the people lead, the leaders will follow." Sooner or later, the truth will come out. But the process of truth-telling appears to be much too slow for us to be able to plan our futures in a timely and rational way.

I'm especially struck by the collective's lie-by-omission about radical innovation in clean energy and other sustainable solutions. Sadly, the lie is shared even by the most progressive scientists, politicians, journalists and environmentalists. To them, there can be no clean energy breakthroughs, period. Only when pressed, these mainstream scientists and pundits claim advanced energy technologies are not even worth researching because "we all know you can't break the laws of thermodynamics." They are wrong.

This is a lie because those laws are broken many, many times under nonequilibrium conditions, for example, by quantum experiments and by research on various devices that show anomalous energy coming from electromagnetic, plasma, solid state, and electrochemical devices. These experiments are systematically debunked by some of the most vocal and powerful physicists who really don't know what they're talking about. But these high priests of a decadent materialistic physics provide a convenient and effective censorship on bold new breakthroughs as the oil barons laugh all the way to the bank and the planet descends into ecological, political and economic tyranny.

The schools of Richard Heinberg (*The Party's Over*), George Monbiot (*Heat*), Michael Ruppert (*Crossing the Rubicon*), James Lovelock (*The Revenge of Gaia*), Tim Flannery (*The Weather Makers*), Ross Gelbspan (*The Boiling Point*), Michael Klare (*Blood and Oil*), Bill McKibben (*The End of Nature*) and countless academic scientists excel at stating the problems of global climate change and dwindling supplies of oil. Their despondency about the lack of viable solutions beyond sacrifice, and our seeming inability to avert imminent global collapse, show on virtually every page of their writings.

But many of these authors can only scoff at considering even the remotest chance we could violate any of the sacrosanct "laws" of physics, and that we must evermore live with existing technologies to obtain our energy. Needless to say, their prognosis is truly grim—as it should be, under the limiting assumptions they themselves have placed on their analyses.

The sad result is an unwitting alliance between the powers-that-be and those partial truth-tellers who are articulate about the crisis itself but are woefully ignorant about the full range of solutions. Layers of truth unfurl at a frustratingly slow pace.

We all lose from this creep of perceived credibility. Greater truths seem to be embraced only in fleeting moments of tiny bite-sized increments, but drown in the cacophony of hubris and kitsch. We are led to believe that our awareness can only take on so much at a time. We are also led to believe that any true energy breakthrough could only occur in the distant future, at best.

As we move through each unfurling layer of illusion-to-truth and its propagation out through the collective consciousness, the Earth clock ticks ever onward towards exhaustion. Our growing biocide and genocide moves much faster than we can respond. This censorship of underlying truth, especially by those who should know better, is but another deception, a lie of omission, which pre-empts rational discussion of the wide variety of choices we really do have.

What we see happening now is that the wave of consensus awareness sweeps through the culture first towards those "solutions" such as biofuels, nuclear power, 'clean coal' and hydrogen cars, which are really nonsolutions in the long run. Upon intelligent examination, none of these alternatives by themselves or even in combination could adequately replace hydrocarbons to meet current demands without serious economic and ecological consequences. Yet they're the only ones allowed to be discussed in the public domain because they represent only what we now know and have a vested interest in.

History teaches us the same dynamics of the denial of new possibilities again and again. Many wise people such as Thomas Jefferson, Abraham Lincoln, Mark Twain, Aldous Huxley, Martin Luther King, Albert Einstein, Franklin D. Roosevelt, John F. Kennedy, Bertrand Russell, Albert Schweitzer, Buckminster Fuller, George Bernard Shaw, Arthur C. Clarke, Margaret Mead, and the whole genre of science fiction point out time after time that new ideas are often not considered seriously, often until it's too late to implement them wisely or peacefully. Out of fear and ignorance, we give our power away to those who are defending an old paradigm that could kill us all. We shut off our imaginations, our vision. We all collaborate in committing lies of omission.

I know about the resistance to discussing real change only too well. Even though my scientific credentials and long research career are impeccable, my presentations on new energy possibilities are most often ignored or debunked by my former academic colleagues, the media and a bewildered general public.

For example, I was recently invited by a BBC producer to talk about new energy on a special documentary on future energy alternatives, but then uninvited when an executive producer decreed that they would cover only nuclear "hot" fusion Tokomak reactors as the one "credible" new energy option. This research, long supported by the scientific establishment, has so far turned up nothing, yet has cost governments tens of billions of dollars. Some of these same scientists have fraudulently debunked, defunded and derailed early experiments by some electrochemists suggesting that clean and cheap energy could come from low temperature catalytic nuclear reactions in water and heavy water solutions ("cold fusion").

The frauds, the lies and the cover-ups can only get worse as the truth marches on, polarizing the culture ever more. We are on a collision course towards an unprecedented revolution in which either the people awaken to deeper truths to be acted upon, or we all perish and bring down nature with us. My new energy colleagues and I keep getting muzzled and debunked because the corporate cartel running the world want more and more control over our energy policy. Meanwhile, I feel that the closer I get to the truth, the further I feel pushed away from the culture.

Once again, censorship has almost totally cut off my access to the media, which I used to enjoy as a mainstream scientist for decades. This is a grievous lie of omission. Hundreds of publishers have also turned me down in spite of a productive and profitable background of authorship. They too lie by omission. So how is the collective mind to ever learn about these possibilities with all these media blackouts? It seems that the only effective answer to this obstacle is to trust in the universal mind or hundredth-monkey phenomenon, that somehow these developments will see the light of day with the consciousnesses and interconnectedness of committed positive intention by those of us dedicated to the responsible deployment of free energy (we return to this in Chapter 21).

To the censors, I've become a heretic who has abandoned his comfortable positions at prestigious universities to be perceived as going off the deep end, of tilting at windmills in exile, one who uses bad science to bolster some extraordinary claims. Worse, I am one of those conspiracy theorists not worth listening to, for surely the existing world-views must prevail for us to mold our future. It's all a Catch-22, a mish-mash of fuzzy thinking, obfuscation, denial, careerism, and self-protection. "Better the devil we know than the devil we don't know, so leave me alone and let me do my work" is the usual refrain.

Progressives, mainstream scientists, environmentalists and Al Gore alike have stopped far too short of telling us how to meet the mandates of the climate crisis and the depletion of natural fuels. My attempts to engage most of these spokespeople about new energy remain unanswered, in spite of broaching the discussions only as a hypothetical possibility, as a what-if. As admirable as their critiques of current energy policy might be, they speak to only part of the truth and so cover up the real solutions.

In the big picture, these people lie by omission almost as much as the oligarchs, and by default, most of the rest of us go along with the propaganda. New energy truth will emerge in time, but we do not have the luxury of time to keep the charade of believing "free energy cannot exist" indefinitely. Our leading scientists and communicators are going to have to stop riding the slow wave of unfolding layers of partial truth in such politically correct ways—or step aside and let others take over. We need to make it safe to accelerate the process of unravelling truth. We need to be willing to suspend disbelief.

If we want to uncover deeper truths we must develop a new perspective and reframe the issues. We must be willing to be perceived as misguided heretics. For years I had a bumper sticker that read, "The truth will set you free but first it will piss you off." That's fine. But then we will need to muster the courage to face the truth with action. We can begin to do that by creating a greater context from which to ask our questions.

The necessary re-framing begins with leaving our prior beliefs at the door (even if only for a moment). We must be willing to have discussions we've never had before. It begins with a simple neutral statement like: "Say we spend a small amount of our collective resources on exploring the possibility of having clean, cheap, decentralized energy for all humankind. Is this worth the effort? If so, how can we implement this research, development and deployment?"
There are at least two important reasons why we should have this discussion.

First, we can begin to address the full range of possibilities for a clean energy future and a sustainable Earth. How can we make informed decisions based on incomplete information? Is there not any acknowledgement of energy innovation? Why is this topic so taboo, especially in the face of such a planetary crisis?

The second reason is just as important: Who will benefit by going to this or that energy choice? A new energy future must be able to benefit all of humankind, all of nature. Therefore, we cannot, *must* not, give this one away to the oligarchy who have benefited so much from controlling nonrenewable resources and destroying our environment.

Those who run the world know this best. They have structured things to suppress authentic solutions. They optimize fear while they optimize profits, wanting to squeeze out every last drop of oil, natural gas, uranium, water, wood, topsoil, crops, coal or anything else we all need until these resources are utterly exhausted. History tells us that if we wait too long, a time might come when they take their last profit and then run for the hills moments before being lynched by a just-awakened mob, or they'll blow us all up in a World War III. And the rest of us, in our silence, are complicit with this madness. In such a case, we lose our freedoms, our environment, our health, our peace.

As a result of our unabated thirst for hydrocarbons under the ground, two visible communities are forming a consensus about the most pressing issues that mandate a drastic reduction in our consumption of oil, gas and coal. One community is made up of competent climate scientists and ecologists who warn us that we must drastically cut back carbon dioxide emissions and deforestation very soon if we are ever to reverse drastic climate change. The second community, made up of oil geologists and economists, warn us that the decreasing supply of available oil and other resources in the face of increasing demands (the "peak oil" movement) will create such international havoc that wars and economic collapse are inevitable—unless we quickly switch to viable energy alternatives.

The irony here is that, those same experts who justifiably warn us of the grim consequences of our actions also deny the full range of possible solutions such as new energy. Meanwhile, some progressives working within the system are slow to respond to the problem and have not the foggiest idea about how to really solve it. Others surely do know but they lie-by-omission to stay in the good graces with their powerful corporate sponsors.

Politics is about the art of what's possible, not the science of what's real. The U.S. and other western capitalist "democracies" are sacrificing the truth on the altar of economic expediency. The accumulation of money and power for the privileged few is what runs the world. These oligarchs perceive that authentic democracy in energy independence as a threat, because they would lose their power if new clean, cheap energy were to become available. We are talking about supplanting a multi-trillion dollar economy that underlies their immense power. That's why they are lying by omission, that's why so many inventors have been threatened and assassinated.

Breakthrough energy can be introduced for the benefit of humankind by a common revolution by the people themselves. But only through a cultural-political process of truth and reconciliation can we reveal the answers with which we must develop a new consensus.

We are entering an era of truth-telling coming from the people, not from any politician or mainstream media "source": 9/11 truth, constitutional truth, electoral fraud truth, war pretext truth, depleted uranium truth, UFO truth, consciousness science truth, and many others. If new energy truth is to take its place on this list, we are going to have to organize ourselves as activists to force the body politic to embrace truth, cutting across many issues and forming alliances. Otherwise, we will surely enter a Dark Age of enormous proportions and most all species will die off from human folly.

Just as important as truth-telling is the reconciliation of our past with our future. The ending of apartheid in South Africa and the falling of the Berlin Wall are modern examples of truth and reconciliation. We must not only serve justice upon those who have knowingly lied for their own benefit and to the detriment of humans and nature, we must reconcile ourselves with our past, with one another and with all creation. As we shall see in Part III, solution energy truth and reconciliation are key elements in these times of great change.

This essay underlies speeches presented in May 2007 at the International Institute of Integral Human Sciences in Montreal, Canada (www.iiihs.org) and on the May PQI Mediterranean cruise Q3 advanced course (www.pqievents.com).

Chapter 8
Not All "Conspiracy Theorists" are Paranoid, Many are Truth-Seekers

"The reasons for (UFO) secrecy are simple: The inertia of highly classified programs, embarrassment over past illegal actions taken to enforce secrecy, and the fact that the energy and propulsion systems behind the mysterious UFO objects have been studied and fully understood. This disclosure would spell the end for oil, gas, coal and other conventional uses of power—and with that, the end of the current oil-based geopolitical order and economy."

– Steven Greer, www.thedisclosureproject.org,
November 2007

"With over $100 billion going into alternative energy investments this year, it is time to pause and ask: What is real alternative energy? Totally ignored in the mainstream (and even alternative) media is the area of advanced electromagnetic systems that tap the energy of the endless Zero Point energy field that is teeming all around us. For decades, inventors and scientists have made advances in this area, only to be ridiculed, ignored, or actively suppressed."

– Steven Greer, The Orion Project, September, 2008

Several years ago, while I was still slowly emerging outside the box of scientific and cultural orthodoxy, a new-scientist colleague began to inform me that the whole world was run by a small shadowy cadre of powerful bankers and secret societies bent on consolidating fascist control and depopulating the planet. I had heard a little about this possibility, but at the time wasn't interested in such a preposterous suggestion; I was too busy with other things. For all I knew then, the U.S. government itself was still somewhat democratic and all-powerful, its dark underbelly being not all that great a concern. I believed that, although the government is deeply flawed, it was basically responsive to the will of the people. Nobody could pull their strings was my belief.

But my friend kept persisting that this clandestine conspiracy was so. I found myself getting increasingly angry with him, so much so that I made a big issue of our disagreement, and it nearly separated us from our collaborations. I could hardly contain my rage and yet I seemed to derive some perverse pleasure in my aggression. Only several years later, after researching the matter myself, I began to realize that my associate was correct, at least to some degree. I apologized to him about my behavior, but by then I was the one who was the conspiracy theorist. He had gone on to other things.

Recently I had dinner with an eminent visionary who felt he had "proved" there was no such thing as the paranormal or energy from the vacuum, that all cause-and-effect was purely materialistic. To the contrary, I had made a close study of these phenomena for decades and written books about it, including rigorous experiments that human intention can alter measurable properties of the material world. So after hearing him out for a long time, I then mildly objected to his narrow interpretation of reality. He responded with outrage, raising his voice saying my claims were based on fraudulent experiments which he had already disproven himself. He threw ever more emotional charges into the conversation, that basically cut off the rational flow of our debate. Even when I tried to invoke quantum theory, that the observer can influence the properties of the observed, his emotional outburst continued, although at least there was some intellectual acknowledgement of these anomalies coming from a well-established "respectable" century-old science and its evolving theory.

This, of course, was not my first encounter with a self-appointed skeptic and will not be my last (e.g., see my book *The Second Coming of Science* and the writings of Winston Wu and Daniel Drasin). But what made my observations more meaningful was that we all seem to have a bias towards our own deeply-held beliefs of the familiar, the safe, and the structured. No topic appears more controversial than the motive to guard the gates of the coveted "laws" of science, which intellectually I knew to be only theories that apply to a limited range of conditions. Having experienced these biases myself, I had learned that, when it comes to paradigms or worldviews, the institutional scientists were among the last to embrace innovative concepts.

Yet the emotional insistence on the conservative position still carries the day, and the skeptics, who seem to be most all of us upon first exposure to an unsettling truth, want to pre-empt the discussion and become the first guardians of accepted reality. So, by default, the truth about solution energy, about the possibility that the World Trade Center towers collapsed by controlled demolitions on 9/11, about the carnage and torture in Iraq or the integrity of the president, for years have been swept under the rug as non-issues in mainstream political and media discourse, until the truth appears so obvious that it is like the proverbial elephant in the living room.

What is it about the truth that can be painful at first blush? A fascinating study by Dr. Drew Weston at Emory University in 2006, using brain-mapping techniques, showed that "we derive pleasure from irrationally sticking with beliefs against evidence" because there are "flares of activity in the brain's pleasure centers when unwelcome information is being rejected." (Barry Zwicker, *Towers of Deception*, 2006). Conversely, compelling new information and startling truths first reach the pain centers of the brain and are therefore commonly rejected. According to Saunders, this is related to the "hardwiring of individual self-preservation, i.e., we strive to preserve ideas that are part of our identity, and reject anything that conflicts with that."

This phenomenon of human nature is a basic tenet of psy-ops and mind control. Truth-seekers who search for contrary evidence to media-fed official consensus perceptions of reality are not only labeled as a "conspiracy theorists," they can also suffer psychological pain in what is clearly an unpopular and uncomfortable endeavor. It is much easier to join the crowd that ridicules and persecutes the messenger.

So what, really, is a conspiracy theory? The Cambridge International Dictionary of English defines it as "the belief that unpleasant things which happen, especially to governments, are planned by people who want to cause difficulties and do not happen by chance." But is a theory merely a belief? In science, a good theory (e.g., relativity and quantum physics) is a model based on the best available evidence, but not yet etched in stone as a law. That happens only when we have enough evidence to definitively prove the matter under a wide variety of conditions.

A good theory is not merely a guess, opinion or hypothesis. Yet the very term "conspiracy theory" itself seems to have evolved into something demeaning. We seem to be dealing with terminology rather than reality in examining what we really mean by conspiracy theory. There is nothing fundamentally wrong about formulating conspiracy theories and testing them against the evidence; but there can be some very good theories and very bad theories, depending on the integrity and rigor of the investigations.

In fact, the scientific method, properly done, is a dynamic relation between well-executed experiments and theories. Competent scientists test their initial hypotheses, which are educated guesses, by experiment, resulting in new hypotheses and new experiments. When later experiments and hypotheses become more and more refined, then it may be possible to develop a model or theory that explains and predicts a scientific principle under a wide range of conditions. Theories can be highly developed.

Most of us remember the comical but tragic character played by Mel Gibson as a New York taxi driver, whose paranoid fantasies ran his life. George W. Bush reportedly watched this film *Conspiracy Theory* with great pleasure, as he indulged himself with his own fantastic "conspiracy fantasies," for example, about his own paranoid megalomaniacal delusions about winning wars, building empires, terrorizing people and destroying the biosphere.

The fact is, the Bush administration, the ones to follow, and most of Congress are mute about the depth of the crimes our leaders are committing against humanity and nature. Why are they doing this? Have they been told on no uncertain terms that if they were to stand up against these "treasonous" actions, they would not only have their campaign funding cut off, they and their families would be assassinated? In this century, the deaths of Paul Wellstone, Mel Carnahan and John F. Kennedy Jr. come to mind, as are the anthrax attacks on the U.S. Democratic leadership just after the 9/11 attacks. Perhaps our policymakers are being silenced at gunpoint. Many new energy inventors have also been silenced at gunpoint or by murder. This is the kind of conspiracy theory that could explain the reticence of those who could really make a difference...and the public and biosphere pay dearly for it.

In his book *Debunking 9/11 Debunking*, the brilliant philosopher David Ray Griffin, has suggested that the "official" story of the 9/11 attacks was itself a conspiracy theory, and not a good one at that. His scholarly analysis is based upon a careful study of massive evidence pointing to the attacks as being an inside job—another conspiracy theory that is logically more consistent with the facts. Obviously, someone, some group of conspirators, had to do it. The official story claims that 19 Arab hijackers under the supervision of Osama bin Laden and al Qaeda carried out the attacks, that nobody associated with our own government had foreknowledge or participation, and the World Trade Center collapsed from fires resulting from the impact of the jets.

Researchers in the 9/11 truth movement are advancing another theory based on extensive physical and circumstantial *prima facie* evidence which could and should be tested in a court of law. Regardless of how you may feel about who did the attacks, it would seem to be logically prudent to level the playing field in this "who-dunit," and to conduct an impartial criminal investigation that would examine all relevant evidence and motives, regardless of our comfort levels in doing so. We need to follow the trail wherever it leads. We have many reasons to believe that the attacks were an inside job and that a new investigation is urgently warranted.

Rather than the mislabeling of the outsiders comprising this movement as mere "conspiracy theorists," Griffin convincingly argues that, upon critical examination, the official story itself is an untenable conspiracy theory. More importantly, he concludes that in lumping all unauthorized attempts to piece together the circumstances of and responsibility for the attacks as bad conspiracy theories, the authorities and media provide a grave disservice in not discerning the more coherent alternative theories from the more whacky ones. Debunking the latter and ignoring the former can only fuel the fires of official orthodoxy.

Understanding the causes and effects of 9/11 become all the more important when we look at the bigger picture. Griffin wrote:

"This (official) story (serves) as a national religious Myth, used to justify two wars, which have caused many (hundreds) of thousands of deaths, (and) to start a more general war on Islam, in which Muslims are considered guilty until proven innocent…(one) destructive conse-quence of the attacks was their use to focus the public and Congressional mind almost exclusively on terrorism, thereby distract-ing it from the ecological crisis, which is arguably the overarching issue of our age. For the first time in history, one species, our own, is on a trajectory which, if not radically altered, will soon bring our planet's life, at least in its higher forms, to an end. The prominent issue of our day, therefore, should be whether human civilization can learn to live in a way that is sustainable…the violence of 9/11, along with the official narrative thereof, distracted our primary attention away from the relation between humanity and nature and forced it back to human-vs.-human issues." (David Ray Griffin, *Christian Faith and the Truth behind 9/11*, p. 183, 2006, and *Debunking 9/11 Debunking*, Olive Branch Press, p. 320, 2007)

We may never know for sure who "did" the supreme crimes of 9/11. But what we do know is that this tragedy can only divert us from uncovering the truth of the potential of breakthrough energy. The facts become buried ever more deeply under layers of disinformation, deception and hubris (more on 9/11 in Chapter 20).

The accusation that outsiders who seek the truth based on the best available evidence are merely "conspiracy theorists" is clearly a mind control exercise which, in Orwellian Newspeak, gives the false impression that anyone who questions official pronouncements is a nut case. Ironically, the authorities themselves can tell the tallest tales of all, poignant sacred myths that opened this violent new century. My own exile from the ranks of mainstream science to the land of conspiracy theorists project an image of someone who has gone off the deep end, just like the taxi driver in the Mel Gibson film.

What I feel is actually happening is that our unwillingness to unveil basic (painful) truths can provide a dramatic counterpoint— one that could sprout the new life of solutions from the fertile bullshit of the massive corruption of a leadership bent on criminally destroying our democracy. They are suppressing our freedom to pursue those realities that can lift us out of the fears of the unknown and of death. Yet embracing these realities could lead us into a truly sustainable future.

The process of change, of forming alliances of truth movements, desperately needs to recruit ever-increasing numbers of people. This initiative should include at least the following steps: (1) seek the truth of major cultural issues and assertions by examining the evidence, free of vested interests, (2) face the well-known psychological pains of truth-telling rather than experience the pleasures being part of the official conspiracy of zealous liars, (3) face the dangers of being ridiculed, threatened, cut off and even assassinated, (4) formulate sensible future policies in which we can truly have a peaceful, sustainable and just future worldwide, and (5) create a nonviolent revolution and concomitant restoration of the spirit of the U.S. Constitution to make that so (Part III).

Chapter 9
Is New Energy the Holy Grail of our Time? Yes, if We're Wise

"There is limitless and practically useful energy anywhere you stand or sit. We can feel it and we can be it. We can tap into it with real technologies. We can light our homes and drive our quiet, clean vehicles as the wounds of our tortured Earth are healed forever. Let's take our planet back. Let's start now."

– Adam Trombly, www.projectearth.com

"I am glad that I have lived long enough to see this! It is simply wonderful! I hope and pray that you live long enough to see the principle upon which this marvelous artifact is based become the new energy source for all of the passengers in Spaceship Earth."

– R. Buckminster Fuller, www.projectearth.com

I ask a hypothetical question: say that a novel energy source which was clean, cheap, renewable and abundant were being researched, with several positive results coming in, but whose underlying theories remain to be well-understood or accepted. Would you chip in a little to push the technology along towards becoming a practical energy source for everyone on Earth? Or would you support a political candidate who endorsed new energy research? If your answer is yes, you may or may not already understand that many such sources are being researched, one or some of which could solve the energy problem once and for all. But the work is being actively suppressed from public discourse.

The global energy crisis has become an epidemic. We cannot any longer rely on worn-out policies which dramatically increase the chances of oil wars, resource depletion, escalating costs, price-gouging by energy companies, blackouts, more toxic air pollution, global climate change, aging fossil fuel plants, unsafe nuclear power plants, unsightly grid systems and outdated internal combustion engines running on oil, earth and food.

Perpetuating these systems poses a grave threat to humanity, the Earth and all living beings. This path is also economically untenable. Fortunately, the new technologies do exist, waiting for their opportunity to enter the mainstream. Unfortunately these technologies have been repressed by an insidious collaboration between the energy industry and the U.S. government and imitated throughout the world. We must design new energy systems that are publicly owned, or at least overseen, for their timely introduction.

The research and development of these technologies need our support. Whether they be cold fusion methods, advanced hydrogen chemistry or zero-point (vacuum) energy, we have the potential for a zero-emission energy economy by 2020, if openly pursued. Some of these energy technologies could give us a quantum leap in cleanliness and low cost.

The myth has spread that, if these developments were real, we'd have them by now. But we are only in a research phase, which traditionally is publicly funded. The myth also says we must await a fully-deployed "magic bullet", which would be like asking the Wright Brothers to deliver a DC-3 or a Boeing 747. This myth is shared by most scientists, environmentalists, and progressives (the SEPs), whose disbelief and lack of education about the alternatives unwittingly feed the very vested interests who want to perpetuate the old energy paradigm.

We must come together in community to fund, discuss and debate new energy policies and options. As a first step, about $100 million given to leading new energy researchers will allow us to move ahead toward a zero emission world by 2020. The urgency of this situation provides an unprecedented opportunity for progressive political candidates to champion novel energy initiatives that will allow the public to once again take charge of their own destinies.

About 80% of the global economy, directly or indirectly, is now vested in outmoded, polluting energy and weapons systems. Those resources could be reallocated to clean energy and towards an Earth Corps to clean up the planet.

Even if the breakthrough energy research were to go nowhere—a highly unlikely scenario—would it not be in our best interest to move ahead anyway? The main reason the research is now moving slowly is that the inventors are largely isolated and under-supported. We have a chicken-and-egg problem in which the prevailing myth continues to be, "If we don't already have it, it can't be real." We cannot any longer blindly follow this myth, or we face our own extinction. More than any other single step, the reallocation of resources under public direction from the energy barons and warmongers towards life-sustaining technologies will go a long way to create a clean, secure and humane future.

Supporting innovative energy research of all kinds would be a significant first step in literally returning power to the people. But we need your help. Your participation will go a long way towards ending the current nightmare of outdated energy technologies and a liberation of humankind from the bonds of misplaced power. We return to how you can do that in Part III.

Moving into a new energy economy is a physical necessity that transcends political questions of public versus private ownership, liberal versus conservative governance, saving existing investments and jobs versus striking into the unknown. As during any large change in science and technology, the dominant scientific opinions (and amplified by the media) resist change, because the status quo is politically correct for careers. Blocking the way also are the vested interests of environmentalists and progressives.

This will be totally new! We are talking about supplanting the largest financial entity of all time: the energy industry. Now we know why the coming energy revolution has been retarded by those few who control and profit from what are unusable nineteenth and twentieth century technologies.

Shouldn't we openly discuss what new energy could do for us? How can we convert our priorities from the current destructive ones to the new ones? Do we have the courage to move into the needed transitions? How can we make all players feel safe, to adjudicate the transition with fairness to all? We need to get smart, to anticipate the changes. This should be a crash program.

I just enjoyed reading Dan Brown's best-selling novel *The DaVinci Code*, certainly a thriller and mind-bender. Without giving away the plot, the book concerns the quest for the Holy Grail, inspiring deeper questions such as, should culture-shattering secrets be revealed? Do we have any business tampering with anything that threatens vested interests? In our hearts we yearn for truth and justice, but the real world can make that shift difficult if not impossible—yet eventually necessary for the greater good, for the upliftment of humanity.

Some researchers have suggested that new energy is the Holy Grail of our time, to replace current energy systems. What we are now doing in the energy sphere makes no sense and won't work. Hydrocarbons in the ground, mostly petroleum, comprise by far the largest market share of raw materials on Earth—about 74 percent of our total energy use. Humans spend a staggering $4 trillion per year for all this, as supplies dwindle and prices rise ever further.

The United States, with 4 percent of the world's population, consumes a quarter of its energy, and more than the entire planet used in 1950. Our thirst for oil motivated the 2003 American invasion of Iraq, with the world's second largest petroleum reserves, themselves now worth over $4 trillion; and rising. These resources have been recklessly plundered as booty to serve the interests of the energy industry.

The quest for oil is a false Holy Grail, mostly rewarding those seeking windfall profits. But the supplies are limited, burning oil pollutes the air with toxic agents, and it produces devastating climate change and the rise of sea levels. Eating the Earth for coal, tar sands and oil shale are even worse for the environment, the attempts to capture and store it in underground caves is a joke.

Insurmountable problems also beset nuclear power. As our sixty-year history of nuclear energy shows, we created a monster of weapons proliferation, safety concerns and radioactive waste. Nuclear energy is not our Holy Grail! Nor are chemical or biological agents. Nor is the ultimate neoconservative vision of "full-spectrum dominance" by seizing the high ground of space with weapons that could destroy anyone or anything on Earth or beyond. More false Holy Grails.

Nor would our reliance on conventional renewables become the Holy Grail: silicon solar photovoltaics and wind turbines are intermittent, diffuse and materials intensive; hydrogen requires expensive infrastructures and consumes more energy to produce than we can get out of it; and biofuels steal food out of hungry mouths yet still throw carbon into the atmosphere. Granted, solar, wind, hydrogen and some limited biofuel systems would be far preferable to what we have now, but why settle for less in the long run if newer, more elegant options could be developed?

I believe that solution energy, if properly applied, could become the Holy Grail of our time. But as in any transformative quest, many feel that the secret shouldn't be revealed until we're wise enough to earn the right to apply it responsibly. The other energy options are so dismal that we must bite the bullet and make the greatest transition in human history, one that rivals the development of agriculture and the industrial revolution. About this prospect, Bucky Fuller wrote:

"I am afraid there could be hell to pay. I do not use such language lightly, I fear that those who are in power will not welcome this wonderful news (free energy). You will need all the help you can get and then some. We will need the full genius of this self-regenerative universe to help us navigate through what could very well become a more and more narrow passage to a benign future...If you can bring forward this beacon of hope then perhaps the abundant and delightful future I know is still possible can actually be realized." (www.projectearth.com)

Fuller's "narrow passage ahead" is upon us now, and the resistance to change is absolutely enormous. Pioneers are dropping off like ducks in the shooting gallery of change. Just as important a question is, "how can we develop these technologies without abusing them?" So far, it has been ignored in the public eye and yet is being quietly researched in the realms of governmental secrecy for potential weapons use. This too is a false Holy Grail, protecting the interests of only a controlling elite and endangering the rest of us. We must therefore require new energy for the greater good to be benign, meaning no weapons use and no overuse and no more suppression.

So far, we cannot guarantee that result without a transparent study and selection of the most benign, clean, cheap, reliable and decentralized sources now being researched. So far, we have only heard from the interests of big business and their governmental cronies and their SEP critics. In search of profit, those in control could only continue to offer false Holy Grails, and the critics are now too scarcity-conscious to lead the way. But for those of us who choose to "look through the telescope," we can see a wide range of solutions to meet our criteria.

It has been often said that the Holy Grail will find you before you find it if you are worthy. Our worthiness in creating a sane and safe energy future must be based on making intelligent choices rather than by persisting with our currently destructive energy policies motivated by greed, and our current focus on too-little, too-late solutions. Why couldn't we re-allocate current policies to new ones, such as developing a second-generation solution energy program and a general planetary cleanup?

If we proceed wisely, we will soon discover that solution energy will benefit us all if we can let go of outdated power structures, who would be the only ones to "lose" during the transition. We will need to form an alliance of movements to seek peaceful methods of making the transition to a just, peaceful and sustainable future.

This chapter is adapted from two posted essays, The Energy Crisis: Global Salvation or More of the Same? (2003) and an essay the same title as this chapter (2004).

Epilogue to Part I: A Friendly Dialogue about Free Energy and its Resistance

Having read through these first chapters, you may feel it's a no-brainer to consider the possibility of developing free energy. Yet we still resist. The following is a recent dialogue I had with a friend, which still reveals a resistance to a serious discussion about this option. This individual was by no means a scientist or plutocrat, but a well-informed layperson who at least acknowledged I might be right. This dialogue was: (1) symptomatic of the hundreds of inquiries I get about some of the half-measures that make it into the news; (2) symptomatic about the assumptions most of us have about any interest or activism about free energy because of our collective cultural nonbelief in its relevance (a vicious cycle); (3) symptomatic about how lonely our work can be as a result of this disinterest; and (4) symptomatic that we need to review how serious the situation is and what options we really do have other than free energy—and it doesn't look good.

Correspondent: How about the air car India is developing?

BOL: This is a cute concept but another stillbirth in the long run, like the electric car or hydrogen fuel cell car. It takes more energy to compress the air than you get out of the car! One needs to look at the full-life-cycle cost of each option, and all conventional ones like this, I regret to say, fall short.

C: I realize that may be true, but if we were to get that energy to fuel these vehicles from other sustainable sources such as wind and solar wouldn't we be further ahead of the curve and begin to bend the worlds minds into accepting the concept of "free energy?"

B: Yes, it's definitely an improvement, and I would have wholeheartedly supported this option if I were to go back to the knowledge, awareness and wisdom about the subject I had had thirty-five years ago when I was an assistant professor of science policy assessment at Hampshire College. In that capacity, I had taught a course on energy options (a position now occupied by Michael Klare, author of *Blood and Oil*). The problem is, solar and wind are not really renewable. Both are materials-and-energy-intensive to extract, manufacture, transport, build and maintain, they are sources that are diffuse and intermittent, and they use up a lot of land in unaesthetic ways. Windmills kill birds and make loud noises. Only free energy is clean and renewable and totally decentralized. It was after two decades of further research, world travels to visit with the inventors, and then writing books and essays, when I became convinced about the credibility and diversity of the breakthrough concepts themselves and the air-tightness of the efforts to suppress them. It took me two decades to be able to gain the confidence to take the quotes off the phrase known as "free energy" or whatever we'd like to call it. I've seen many demonstrations for myself. Meanwhile, I hope it won't take as long for others to understand the potential.

C: I think the problem with the concept of "free energy" is that it sets off an emotional trigger that people just don't find believable.

B: You have identified the number one barrier to the implementation of a truly clean, renewable energy future, one which promises us all true abundance. There are few souls who are even willing to address the question of free energy, regardless of how they may feel about it or how they think the public might feel about it. Meanwhile, most of us are quite literally tilting at windmills searching for old and familiar solutions, analogous to the discussions about aviation between 1903 and about 1908, when the Wright brothers were already flying. Talk then was almost entirely about what kinds of balloons, airships or dirigibles could fly passengers. *Scientific American* ran an editorial in 1905 stating that the Wrights were a fraud.

C: We are conditioned that there are no "free lunches." In order to improve the communication stumbling block I think you need to rethink the approach so that the "masses" can begin to embrace the concept that energy can be developed, sustained, and affordable without being held hostage by oil.

B: Yes indeed, that's my job now, and it's a very big one, and I need help, even if it's just moral support. My colleague Wade Frazier recently addressed the issue that we'd almost prefer to end civilization than to open ourselves to true energy freedom (www.ahealedplanet.net/paradigm.htm). Shortly before their deaths, the famous visionaries Buckminster Fuller and Arthur C. Clarke both became aware of, expressed, and supported the enormous potential of an abundant civilization powered by a free energy economy. Embracing this will require a paradigm shift that transcends the cultural biases which Frazier identified as nationalism, capitalism, scientism and structuralism. I describe these blockages collectively as "pragmatism" or conventional wisdom (Chapter 18).

C: The world stills embraces the concept that oil is the only reliable and affordable energy available to us. I hear this all the time on reports from NPR (National Public Radio) and other writings. Until we change the mind set oil will be the fallback energy source.

B: Absolutely. Shortly before we made the decision to leave the U.S., I was once listening to NPR and heard this: "The following program is being brought to you by the Rockefeller brothers, where sustainability is a way of life." I was so disgusted by this Greenwash that I turned off the radio permanently. It had the effect of accelerating our decision to get out. So much for *public* media, it's all corrupted.

C: I think the recent inflation of oil on the world market is opening up huge opportunities for more sustainable energy and conservation methods. Mass transit is on the rise. People are looking back at the old ways of farming that were sustainable but lost in the frenzy of the chemical charade.

B: Yes! The latest in our global oil glut was the recent (suppressed) news that 180 mostly contiguous "blocks" for oil-and-gas-drilling in the Amazon just 100-500 miles to the East of us, about the size of Texas, are being leased to over 35 multinational companies. This is insane! Drilling in such a vast area would virtually assure the obliteration of the largest, most biodiverse rainforest and carbon sink in the world, including dozens of voluntarily-isolated indigenous peoples whose contact with Westerners would give them diseases for which they have no immunity. Now there's the potential for desertification of this entire region. Resisting these projects and going for the alternatives are absolutely necessary.

C: We are in an era of great opportunities and I think your work on energy needs to continue in a "big" way. Acceptance may not come immediately but when the rosebud is ready to blossom you will have the kindling ready to stoke the fires.

B: Thank you for acknowledging this, but the work needs to be understood by many more of us in the advance guard; otherwise Big Oil will keep winning by default. You and anyone else who have progressed as far as you have are a pivotal part of doing this job; I am just one person who happens to have years of knowledge about this option and who can help people learn what is possible if only more of us can keep watering the rose plant. This is a collective effort, both articulating the problem and understanding that any sane future energy policy needs to leave no stone unturned in our quest for truly sustainable energy. One of my main challenges these days is to encourage people to look at the REAL breakthrough (suppressed) solutions in the midst of all this commercial hype about half-measures, which although symbolically important, can only make way for what can be truly lasting. The bad news is, we don't have much time, it's running out. The good news is, free energy needn't be that far off in time nor must the research and development be that costly. The paradigm shift requires a lot of education for those of us ready to take on the mantle of responsibility, and this might include you. Meanwhile, I need all the support I can get on this rather than the usual brush-off, to have the opportunity to teach about the possibilities, to be able to encourage people to suspend disbelief, if only for a short while, and to look at what a world of true abundance might look like. We can only try even if these ideas are not yet "credible" in mainstream and progressive circles.

Part II

The Mandate for Change in Energy Policy:

What Options Do We Have?

Chapter 10
The Moon-to-Moon Lunacy of the Climate Skeptics

"Given the urgency and magnitude of the escalating pace of climate change, the only hope lies in a rapid and unprecedented mobilization of humanity around this issue…The ultimate hope is that—especially given the centrality of energy to our modern lives—a meaningful solution to the climate crisis could potentially be the beginning of a much larger transformation of our social and economic dynamics."
– Ross Gelbspan, *Boiling Point* (Basic Books, 2004)

I live in the Andes of Southern Ecuador. In 2007, President Rafael Correa proposed that the international community match funds with the country to keep the oil in the ground on 1 million hectares of pristine rainforest in Yasuni National Park. This is Ecuador's largest untapped oil reserve, comprising almost 1 billion barrels of heavy crude oil, worth $10-$30 billion at today's prices—but with the potential to destroy a whole ecosystem.

The Save Yasuni effort pales before looking at the situation over the entire Western Amazon region just to the East of us here. According to a 2008 study summarized in the link below, oil exploration, drilling and road-building leases have been granted to over two-thirds of both the Ecuadorian and Peruvian Amazon, an area nearly 100 times vaster than the Yasuni's controversial oil-leased blocks and larger than all of Ecuador itself: http://www.commondreams.org/archive/2008/08/13/10973/

This excellent study, entitled OIL AND GAS PROJECTS IN THE WESTERN AMAZON, can be downloaded at http://www.saveamericasforests.org/WesternAmazon/index.html

We here in Vilcabamba have a constant supply of rain from one of the richest, most biodiverse, carbon-sequestering lungs of the

Earth—the Western Amazon—including some of the few remaining voluntarily-isolated indigenous peoples anywhere. The desertification of vast areas of the Western Amazon would be virtually assured if these regions were to be drilled and roads built haphazardly by the more than 35 multinational oil-gas companies to whom the leases have been or are being granted.

The rainforest is priceless; the oil, however, at today's price, could be worth over $1 trillion. Yet the rare herbal medicines found only in this region could turn out to be worth more. Now they all face extinction along with many other species of mammals, birds and plants found anywhere on Earth. The implications are huge, whether looked at from a NIMBY or global point of view, it can't help but want us to take action. Not only should we insist on a moratorium on all exploration and drilling here, we must bring to public attention the possibility of rapidly developing clean breakthrough energy sources that would render hydrocarbons obsolete. We must keep the oil in the ground before it's too late. The economy will have to change—or we're doomed. I'm not kidding about this.

The scientific evidence is overwhelming that the emissions of carbon dioxide and other greenhouse gases coming from billions of vehicles, powerplants, buildings, fires and tree-cutting are changing our climate. The evidence is also mounting that secret attempts to mitigate these effects by spewing particles (chemtrails) into the atmosphere, sometimes called global dimming or cooling, can destabilize the climate even more. These actions, plus the rampant deforestation of vast regions, can only create more chaos.

In spite of this, some academic scientist-skeptics and some progressives believe the causes are natural and they vocally dispute the veracity of "anthropogenic (human-caused) climate change." My own view is that, while we don't know the exact balance of human and natural causes to global warming and climate change, I am convinced, as an atmospheric scientist myself, and in agreement with the vast majority of competent climate scientists, that the human component dominates the mix. Even if it were not the case, what sane person would still want to pollute and "uglify" the environment?

The models that predict the climatic effects of measured emissions are mostly conservative, and very consistent with the well-tracked observations of the effects, such as melting of polar ice, sea level rise, superstorms, superdroughts and hot weather. At the very least, this needs tracking and that we adopt the precautionary principle and move forthwith towards second-generation renewable energy and responsible government. No matter the outcome of the climate debate, it is clear that burning hydrocarbons pollutes, causes hundreds of millions of lung disease cases, and it creates wars. That's reason enough to make the needed changes.

Because of the propaganda of the climate skeptics, many of my most beloved friends have come to me with the proposal that human-caused climate change is a fraud. This assertion flies in the face of the overwhelming scientific consensus to the contrary and clearly invites in the precautionary principle, which reasons that no matter what the actual situation might be, is it not wise to hedge your bets in favor of considering the *possibility* that the consensus is correct, and therefore we should take all preventative measures because of the clearly present dangers.

In spite of polarized attitudes about the degree of human intervention, the logical conclusion is that carbon emissions should be prevented or curbed at the source, or better still, the fuels should be kept in the ground. But those in charge of our global economy and politics continue to harangue us with distractions toward "solutions" vested in the marketplace (carbon-trading) or not at all, or as we shall see, proposing absurd bandaids such as carbon sequestration and biofuels.

The December 2007 UN climate conference in Bali is a prime example of how distorted the discussion of the extent of our unfolding disaster really is, its causes and its remedies. The December 13 "Open Letter to the Secretary General of the United Nations," signed by eighty-odd climate-skeptic mostly U.S.-based and –biased "professwhores," is only the most recent case in point. This appalling letter leaves us the clear impression that there is no human contribution to global warming or climate change, that it's all part of a natural cycle, and that we should go on "promoting economic growth and wealth generation" rather than joining the rest of the world in cutting back emissions. The authors of this short-sighted and suicidal attitude completely ignore the possibility that we could have any meaningful conversion to a clean energy economy.

Such assertions contradict the well-researched conservative calculations made over the past decade of some 2500 scientists on the UN Intergovernmental Panel on Climate Change (IPCC), whose conclusions are just that opposite: that humans are the dominant component of global warming and climate change. In other words, greenhouse gas emissions rather than "natural cycles" are the major cause. This is concurred by leading atmospheric scientists such as James Hansen, Director of the NASA Goddard Institute for Space Studies.

He and others, who have concluded that the IPCC reports are too cautious, and that we have less than ten years to act radically.

Hansen declared that the U.S. and other Western governments are beholden to the coal industry and the oil industry: "They put out disinformation, they fund a small number of scientists and they expect the media to give you a balanced story. And by 'balanced' they mean that (if most all) scientists are saying that something's happening, it would have to be balanced by someone saying, 'Oh, this is just natural.' ...even though the story has become very clear...it's 99.99 percent certain that humans are influencing the climate—but still, they make the story appear much less certain than it is, and therefore, why should we take actions as long as it's uncertain?" (interview with Amy Goodman, Democracy Now! March 24, 2008)

Even short of the dire conclusions of these scientists, who directly contradict the absurd polemic of the climate skeptics, the well-measured human-caused loading of carbon dioxide (and other polluting gases) into the Earth's atmosphere and oceans are creating a massive biocide that is beyond dispute.

But what's the truth about climate change? This polarization among scientists is quite remarkable and leaves the public bewildered about who's right or who's wrong. The climate debate and resulting inaction also distract us from the larger issues of humanity's war against nature. Nevertheless, both scientific logic and seeing where all the money (bribes) goes, make it clear that the evidence overwhelmingly supports the human-intervention theory for a wide range of effects that are undeniable. In an of itself, going to war to satisfy our oil addiction is outrageous. To summarize the data:

First, it is clear that our combustion of hydrocarbons has so far added more than thirty percent carbon dioxide into the atmosphere over the past century, most of it coming in the past three decades. These actions, combined with deforestation, are overburdening the entire biosphere with excess CO_2. For example, the oceans are unable to take in all the greater amount of carbon in natural ways, and so are becoming more acidified, killing sea-life and coral reefs.

Secondly, the positive feedback loops between increasing carbon-loading and other factors such as polar ice melts (causing sea level rise, darker ground and water uncovered, causing greater absorption of sunlight and even warmer temperatures), ocean and air current changes, deforestation, the emissions of other chemicals, ozone depletion, and their influence on climate change itself, are evident when examined objectively.

Thirdly, whatever the degree of human intervention is happening, we have to do something radical about the situation ("radical" meaning "going to the roots"). This is just common sense if there is to be any compassion we may still have for life on Earth.

The climate debate is such a distraction from the scientific facts that our climate is changing, humans are increasingly destroying the biosphere through the careless pollution, geoengineering, and exploitation of its dwindling natural resources, spending trillions of dollars on the machinery of war, and committing genocide to covet those resources. At the root of the problem lay our destructive energy practices that exploit what's left of the world's oil, coal, gas, trees, food, water and biodiversity. This is the insanity we need to confront!

For example, Kevin Watkins of Oxford University and lead author of a November 2007 UN Human Development Report, concluded, "We could be on the verge of seeing human development reverse for the first time in 30 years...The message for Bali is the world cannot afford to wait, it has less than a decade to change course." (Reuters, November 27, 2007).

Nevertheless, during autumn 2007, I was invited to appear on the Kevin Smith radio talk show, in which for the first hour, Mr. Smith grilled me that, according to a recent neoconservative think-tank Hudson Institute report, human-caused global warming and climate change are a hoax! During our interview, we could never get beyond being mired in such arguments. We didn't even begin to embrace any solutions, or to be able to imagine the world in a post-hydrocarbon age.

One of the most vocal signatories to this appalling "Open Letter" to UN Secretary-General Ban ki-Moon was S. Fred Singer, an atmospheric physicist who had been one of my mentors during my graduate school years during the 1960s, and a collaborator on various scientific projects during the 1970s and 1980s. Over the past two decades, Singer took a curiously abrupt turn to right-wing politics. Moving to Washington, he soon joined neoconservative and big-business lobby groups such as the Heritage Foundation, the Global Climate Coalition, and the billionaire Reverend Sun Moon's scientific advisory panel. Before the climate became an issue, about twenty years ago, he invited me to two of their soirees, flying me into Houston and London first-class. (This may have been another example of a carrot dangled before me to "join the club" rather than reject it.) On these occasions, I began to discover that, simply by following the money, we could really determine where establishment science would go: in service to those who run the world. Like many of his climate-skeptic colleagues, Singer had sold out in promoting this "Moon-to-Moon" lunacy.

So what came of the long-awaited UN Bali meeting? It became a circus in which all nations opposed the U.S.-dominated refusal to support even the modest Kyoto protocols limiting carbon dioxide emissions. The United States, with four percent of the world's population, emits a quarter of the world's greenhouse gases, by far the highest per-capita rate of any nation in the world. It was no surprise that the U.S. representative to the Bali climate talks was booed off the stage for American inaction.

As the tension between other countries and the U.S. further mounted, a delegate from Papua, New Guinea told the Americans attending the conference: "We ask for your leadership and we seek your leadership. If you are not willing to lead, please get out of the way."

The criminal leadership of my native country again stalled meaningful talks and instead proposed limiting emissions by trading carbon and using similar neoliberal market approaches to solve the climate crisis, as if only the elite could eventually afford to burn carbon compounds. This is akin to moving the mess from one side of the room to under someone else's bed. Nevertheless, our "leaders" have asserted that the "solutions" to climate change and deforestation should be managed by...the World Bank!!

The current president of the World Bank appointed by George Bush is the neoconservative, neoliberal Bush loyalist Robert Zoellick, whose most memorable quote is, "Negotiating a free trade deal with the United States is not something one has the right to do...it's a privilege." The World Bank was chartered to end world poverty but instead increased it through its draconian policies supporting the American-and-corporate empire's insatiable thirst for unrenewable resources. It's ironic that the previous two presidents of the World Bank were James Wolfensohn and Paul Wolfowitz. In storybook fashion, the plutocratic wolves keep guarding this chicken coop we call Earth.

What, in fact, is really happening is that the likes of the World Bank loan money to the nations of the South for "development" but many of these dollars are squandered by corrupt politicians who accept a debt-and-interest burden so high, many countries cannot pay it back and sink ever deeper into poverty. They then allow their people, forests, hydrocarbons and metals to be ever more plundered, especially as their value keeps going up. This vicious cycle keeps expanding, usually regardless of political changes.

I live in one such country, Ecuador. This resource-rich nation, blessed with awesome natural beauty, owes the World Bank $10.5 billion, a debt that might be later reduced because some of the loan might have been illegal. Ten billion dollars also happens to be the approximate minimum value to Ecuador of the oil under the ground at Yasuni National Park. Why couldn't the World Bank forgive the debt and give this country a chance to move forward? After all, all loans are just a fictitious digital sleight of hand, not backed by real wealth. And why couldn't Chevron pay the $6 billion of estimated damages for the biocidal mess they left in Ecuador, several times greater than the infamous Exxon-Valdez oil spill, and which they're trying to fight in court? The battle is now joined between the forces that want to exploit and those that want to preserve, and there's no greater example than that of Ecuador.

The truth is, regardless of where we might stand on the Moon-to-Moon lunacy scale, we will have no future without discarding false economic presuppositions that demand the genocide and rape of the biosphere for the last little bits of Earth's resources while suppressing what we must do to survive the crisis.

It is evident that Zionists and some New World Order elites are warmongering and exploiting Middle Eastern oil in order to consolidate their control, blaming the likes of Al Gore and the climate scientists as co-conspirators. Other elitists posing as climate-change alarmists want to push nuclear power and biofuels as the answers, another corporate ruse. These dangerous options can only distract us from organizing a clean energy future when there are so many strong, overlapping and yet contradictory views. We are left with a tower of babble and need to act on climate change, whatever the extent of human input and whatever conspiracies underlie such claims. We must organize ourselves to solve this problem—and free energy provides the answer. Neither blind obedience nor anarchy will do.

Among the few who care or dare to look at the new energy option, some argue that any new or decentralized energy source would be bad for (big) business, that it would stifle economic growth which depends so much on the continuity and expansion of the existing energy sector, to which we are so addicted, hence the Moon-to-Moon lunacy. The new option may actually promote a new era of intellectual expansion, providing it's not side-tracked into more of the same kind of corporate control. True, we are talking about supplanting the first and only multitrillion dollar annual spending enterprise in human history, we are talking about awakening to a new economic agenda on a global level.

Chapter 11
Climate Crisis Call to Action

"Much as I love the green movement, in the face of the countless billions still to be made by raping the planet for oil, they're merely the equivalent of trying to water the rainforest with an eyedropper."
– Mark Morford, *The San Francisco Chronicle,*
Nov. 28, 2007

"Some day this century, the time will arrive when the human influence on the climate will overwhelm all natural factors…then, the judiciary will be faced with apportioning guilt and responsibility for human action resulting in the new climate. And that, I think, will change everything."
– Tim Flannery, *The Weather Makers,* 2006, p. 284

I write this on Easter weekend, a time of reflection on the possible crucifixion of our beloved Gaia and the haunting uncertainty about whether or not we can reverse our destructive practices and help her come to life again. It also comes at a time when our understanding of the context in which we are operating becomes muddied by the exorbitant claims of those who deny that there is any human interference at all with our climate, and that, for the sake of economic growth, we should do nothing about our carbon dioxide emissions, that we should keep going to war to secure dwindling supplies of oil, even though the Pentagon spends hundreds of times more on war and weapons than we invest on mitigating climate change and conventional renewable energy. Zero public funds go to new energy research. What reckless nonsense!

Fortunately, I am joined by millions of highly educated persons in making my protest. Unfortunately, I am a voice in the wilderness when it comes to creating an energy solution revolution.

The intention of this book is to explore the potential of *new,* clean, cheap, decentralized (and suppressed) energy sources to solve the problem. By new energy, we mean energy from the vacuum ("zero-point"), cold fusion, advanced hydrogen and water chemistries and any other technology which could provide a breakthrough, whether acknowledged or not acknowledged by mainstream science. A small R&D effort might be all that is needed to ignite a new energy age.

Out of self-interest, politicians and big businesses are doing everything they can to avoid the real solutions that could save our planet from a climate catastrophe. Yet we have the innovative solutions if only we look and act!

Regarding our voracious squandering of hydrocarbons, "clean coal" is not clean. Nor is oil and natural gas. Nor will supplies last very long. Nor will carbon-trading that grants the elite the right to pollute. Nor will it help much to sequester carbon dioxide emissions into big subterranean domes or to throw particles into the atmosphere to try to offset greenhouse emissions or sprinkle iron into the oceans to absorb more carbon dioxide, resulting in ever-increasing acidity that threatens sea life. Terror-forming Gaia through such gross macroengineering manipulations would only forestall our day of reckoning at the table of her Last Supper, of *our* Last Supper.

It's too late to go for band-aids. Climate scientists are telling us that we must curb greenhouse gas emissions up to 90%, maybe more, by 2050 to even begin to reverse climate change. About 87% of all our worldwide energy use comes from incinerating hydrocarbon fuels. That practice and deforestation comprise almost 100% of the carbon dioxide input into our changing atmosphere, a statistic that even the climate skeptic "professwhores" will have to reckon with.

We cannot go on any longer as we have and continue to cave in to the desires and rationalizations of those who stand to benefit the most or speak the loudest. Our priorities are grossly askew. Some climate scientists give us a maximum of ten years to take drastic action. Do we not want to overcome all the posturing and inertia, and demand that we exercise the precautionary principle? Could we not base our actions on scientific truth?

The first obvious step towards averting climate catastrophe is to rapidly phase out hydrocarbon (fossil fuel) use and replace them with more efficient and cleaner energy sources drawn from a wide range of possibilities. But don't be deceived by what is now being proposed as replacements.

For example, the recent biofuels hype represents a nonsolution. Harvesting and incinerating biomass are destroying rainforests. They bitterly compete with agriculture, increase poverty and famine in the South, and, in the end, inject just as much as or even more CO2 into the atmosphere through slash-and-burn deforestation and desertifying the Earth's greatest carbon sinks. In increasing amounts, biofuel farms are wiping out forests and species in Indonesia, Malaysia, Brazil and many other countries. A biofuel infrastructure also significantly depends on hydrocarbons. Most obscene is the statistic that one SUV tankful of biofuel could feed a hungry person for one year. This is genocide!

Bush's and General Motors's much-touted hydrogen fuel cells are neither clean nor cheap. If the hydrogen were to come from water, it takes more energy to cleanly produce the hydrogen than you get out of it! The only alternative way of producing hydrogen is by extracting it from...hydrocarbons! Nor would hybrid or electric cars or air cars be the answer as long as the electricity or compressed air running them ultimately comes from polluting powerplants. These are all stillbirths when it comes to supplanting the oil, coal, gas and nuclear economies.

The also-touted nuclear power renaissance should never happen: it's too expensive, too vulnerable, too centralized, too radioactive, too much doomsday weapons potential. We don't even know how to dispose of the high-level radioactive waste, which we must leave to untold future generations! So the only choices presented to the public seem to be petro-fascism, nuclear-fascism, or inappropriate-fuels-fascism.

On the other hand, conventional renewables like existing solar energy, hydropower and wind power are getting largely ignored. They're mostly old technologies whose contribution is only 7% of the global total (see the Appendix). There is little room for growth in the present atmosphere (for example, the U.S. Dept. of Energy has projected no change to that modest share through 2030 while we burn more carbon fuels than ever). The U.S. National Renewable Energy Labs stumble along on a measly annual budget of $200 million for conventional renewables. The Department of Energy's projections for 2030 show a steadily increasing use of hydrocarbons and nuclear power and no increase in the proportional share of renewables, known or unknown. They are living a fantasy.

Granted, solar and wind will never provide the long-term answers. Only breakthroughs in vacuum energy, cold fusion, advanced hydrogen technologies could solve the climate problem very quickly. However, researching these technologies need our support if we are ever able to have practical devices.

Our avoidance of the real issues sometimes reminds me of self-important medieval clerics pondering how many angels could fit on the head of a pin or which witches and heretics to burn, while the world enters the darkness of plague and violence. Or Nero fiddling while Rome burns. Or King Louis and Marie Antoinette feeling all is well in their palace in the face of an imminent revolution. Or the stunned staff of the early-twentieth-century aristocrats rearranging the deck chairs of the *Titanic*. Or Hitler in his final bunker plotting his next invasion.

Or the denials of the impending "inconvenient truths" of our own perilous situation. Under media and political fire, Al Gore has well-articulated the climate crisis. But there is a second, more politically incorrect question to ask: do we have the courage to act on *real* solutions? As almost always, most establishment scientists, in order to preserve their grants and careers, deny the truth of new energy.

As a result, the scientists who should know better form an unwitting alliance with the polluters, in spite of abundant evidence to the contrary. But, for the sake of discussion, let's look at the unlikely premise that the skeptics are right, that the treasured "laws" of an existing materialistic physics cannot *ever* again be broken, that "perpetual motion" is perpetually poppycock, and that today's technologies or minor modifications of them are all that is available. Shouldn't we still give innovative energy research a try? Or are these scientists too afraid to be cast among the heretics? Are they too vain to admit they might be wrong?

New energy heresy now eclipses human-caused climate crisis heresy, based on my experience to date. I have seen many successful new energy experiments worldwide, but these courageous pioneers get next to no support; in fact many have been threatened, assassinated or have had their equipment confiscated or destroyed. The denials of new energy are based on the *appearance* of nonexistence rather than its reality, and this social dynamic of non-credibility of the new has been with us throughout history.

The climate scientists have gone out on a limb to make their point. Finally, they are being heard. New energy requires going out on that limb even further, but its consideration must also be taken seriously in the face of our global crisis. But this time, we don't have years to gradually build a consensus.

By default, the leading spokespeople answer *no,* there can be no significant role for outside-the-box energy innovation in the near or more distant future. This paralysis of paradigm is one of the greatest betrayals, whether by commission or by omission. Neither existing "science" nor "economics" can admit of the possibility of developing really good and lasting solutions to the energy-climate crisis. So again we have given our power and destiny away to those institutions that are literally destroying us. In the broader sense, we are all responsible for contributing to humanity's war on nature.

Most importantly, we must put the power of policy and planning back into the hands of the global commons rather than the likes of U.S. vice president Dick Cheney's infamous secret energy task force that loved to hatch wars for oil and line the pockets of big business. Short of replacing the oil warriors, a Democratic-Party-style business-as-usual approach won't work either. We are on the threshold of the biggest change or disaster in recorded human history because of human neglect and greed.

So the powers-that-be resist the new paradigm solutions at every turn, because of the enormous pressure exerted by the energy-auto-financial-media-military complex through their political contributions—especially in the U.S. The politicians are not only unduly influenced by these monstrous human creations, they express (or feign) ignorance of the real solutions. No one wants to take responsibility to manage the pains of transition. They are not even close to addressing the answers that are still politically incorrect but scientifically the only way to go.

When U.S Rep. Dennis Kucinich first expressed interest in drafting a new energy bill and then declared it must be "technology-neutral." I once again gave up on Washington and its numerous vested interests. I support him in his other policies but this one may have been too hot for him to handle. Politics is about the art of the possible within a context of consensual reality mostly among elites, not of an urgent physical or social reality.

If new and innovative energy can get on the agenda only on the coattails of the existing renewables, so be it. But to deny and ignore it entirely could lead to global catastrophe. As of early 2008, no bill yet introduced to a climate-crisis-conscious U.S. Congress or California legislature so much as includes *any* mention or support for new energy research. In fact, the U.S. Patent Office expressly prohibits new energy patent applications as if this research were the thinking of crackpot heretics unfit to participate in the public process.

It seems the only sensible thing that we can do in the short-term is to:

1. impeach and bring to justice those leaders who are committing mass genocide, ecocide, and human rights violations while obstructing those solutions we so desperately need;

2. tax the living daylights out of the polluting Goliath corporations (in a revenue-neutral way), following the partial successes of Europe;

3. support the necessary R&D for clean innovative energy of all kinds Apollo-style, leaving no stone unturned (R&D investments of less than one day of spending at the Pentagon or a few hours of incinerating hydrocarbons could get us going);

4. come together as a global community to embrace the truth, reconcile ourselves with ourselves and with nature; and

5. develop an action plan for transition to a carbon-free, nonpolluting, peace economy and a sustainable environment.

Who in the public spotlight is willing to take this stand? Or could we not create new spotlights for more independent, realistic and visionary thinkers willing to stand up to the polluters? Shouldn't we bring in new blood whom we could truly trust to lead us to make the needed changes? Could we not do this while restoring the integrity of constitutional and international law?

The most important point here is that we have a physically real problem that demands physically real solutions…and so far, few of us are able to express these solutions without being drowned out by the chorus of distracting hubris presented to us by those in power and by loud and ignorant critics.

To do little or nothing about global warming/climate change or to deny its existence because of a sacred mandate to "promote economic growth and wealth generation" is an admission that doing anything significant about the problem threatens the very survival of our hydrocarbon-based industrial economy. This paralysis of paradigm also mandates the destruction of the biosphere, so those of us with the most power and money keep hanging on to the sinking *Titanic* and are more willing to sink than swim. More than ever, we need an energy solution revolution…and soon!

What it all comes down to, dear reader, is that we seem to have but two extreme approaches to our future: (1) stoke the voracious appetite of a military-industrial elite, intent on waging war on nature and ourselves, thereby destroying our environment and our freedoms, or (2) develop clean energy and sustainable biospheric practices under responsible democratic control. Which will it be? Will you join us in stepping up to the challenge?

Some of the material for this chapter is based on an essay of the same title posted on April 7, 2007 and updated in 2008.

Chapter 12
An Open Letter to Al Gore

"At the beginning of 2008, we're chugging out more CO_2 emissions than ever; the world climate system has started to tip over, and it is on record that the USA torpedoed Bali—a policy decision with historic consequences, which isolates America, and which means that U.S. citizens are now bearing a huge burden of guilt."

– Martin Schonfeld, www.commondreams.org
Jan. 7, 2008

Nobody in the public eye and media access more prominently represents the voice of carbon emissions reductions than former U.S. Vice President Al Gore. He has proposed a 90% reduction for industrialized nations and an 80% reduction for non-industrialized nations by 2050. I published and sent to him through mutual contacts the below open letter two years before this writing. So far I've received no response. Will Mr. Gore be politically able to bite the bullet and take his crusade one step further: towards a world of true new energy? Or does he have a gun pointed at his head too to stay silent about the *real* solutions?

Dear Mr. Gore,

I am a former astronaut, Cornell professor, physics faculty member at Princeton University and visiting faculty member in technology assessment at the University of California Berkeley School of Law, I was Mo Udall's energy advisor and speechwriter during his 1975 Presidential campaign, author, AAAS Fellow, World Innovation Foundation Fellow, NASA group achievement award recipient, and founder of the New Energy Movement.

You have asked the public to address the important question, "How can we reverse global climate change?" I agree that taking on that task is critical for our collective survival. You have also stated that we must freeze and drastically reduce our carbon emissions. I agree.

The most promising answer to your question is surprisingly simple and can be summed up in two words: new energy. My experience finds that serious discussion of new energy is still politically incorrect in mainstream circles, which is appalling. Delays in implementing life-saving innovation will be at our collective risk and peril. The urgency for action in these times is unprecedented in human history. Quantum leaps in energy innovation, which some of us in the scientific community are aware of, can provide the needed solution, hopefully in time to avert global disaster.

Having held professorships in the physical sciences and energy policy at many universities with an impeccable publication record for 45 years, I join you in not taking these matters lightly. I make no claims that cannot be rigorously backed up and I have no vested interest in which specific energy options should be implemented. I receive no money for the grassroots work I am doing in assessing these technologies. I can assure you that with proper public support, we will soon have robust solutions without needing many building blocks or wedges. Incremental approaches, as you correctly point out, will not be adequate to solve the problem. But you may not be fully aware of what's on the horizon, since we have been so blinded by our collective shortsightedness.

By "new energy" I mean innovative technologies with the potential of providing a quantum leap in our ability to tap cheap, clean and decentralized energy for producing fuels and electricity. These may or may not be recognized by mainstream science. The technologies include:

ADVANCED HYDROGEN TECHNOLOGIES (1) catalytic water molecule manipulation and dissociation through cheap electrolysis, and (2) manipulation of hydrogen plasmas with catalysts to induce fractional quantum electronic states that yield large energy outputs;

COLD FUSION or lattice-assisted nuclear reactions (LANR) by electrochemical means, induced in water and heavy water solutions catalyzed by (1) palladium cathodes, (2) sonocavitation and (3) other processes that can produce large amounts of thermal, radiation-free nuclear energy;

VACUUM ENERGY or zero-point energy, tapping the enormous quantum potential of every point in space-time, through the use of (1) super-motors with super-magnets (cf. Faraday), (2) solid state devices, (3) Tesla coils, and (4) charge clusters; and

THERMAL ENERGY from the environment.

Any one of the above approaches to new energy promises a quantum leap, i.e., orders of magnitude increase, in our ability to tap and have abundant clean, cheap, decentralized energy for all of humanity. In addition, there are many important transitional technologies which can mitigate emissions in the very near future, as follows:

RECYLING AND SEQUESTERING CO2 AND OTHER POLLUTANTS AT THE SOURCE through innovative chemistry; and

REMEDIATION OF RADIOACTIVE NUCLEAR WASTE with innovative technologies.

All of the above concepts have already been demonstrated in laboratories throughout the world (I have seen many such demonstrations). The results have been published in the peer-reviewed literature, but implementing these technologies has proven difficult because there is no significant support.

As you undoubtedly already know, the environmental literature nowadays well expresses the energy problem and other aspects of our national crisis, but has so far fallen short on solutions. Some of the best scientists in the world (John Holdren, Nathan Lewis, Richard Heinberg, James Lovelock and Ruggero Santilli, for example) have concluded that conventional renewables such as solar, wind, hydro-electric, geothermal, tides, biofuels and hydrogen fuel cells are not nearly adequate to meet current, much less projected, energy demands. Each of these "building block" options runs into serious pitfalls ecologically and economically when we talk about supplanting our multi-trillion dollar hydrocarbon energy economy. Nuclear options also have their serious problems, as I am sure you are aware.

You hit on the situation in your recent NYU speech when you said, "I am certain that some of the most powerful solutions will lie beyond our current categories of building blocks or wedges." You said that America, and only America, has the "capacity for vision" but that "we have to urgently expand the limits of what is politically possible." Very well said, and part of any program to implement new energy will involve a very rapid but necessary political education and risk-taking that even the liberal and progressive community has ignored. I acknowledge, and I am sure you would agree, that the limits of what is politically possible need to stretch very far to accommodate the reality of new energy. But what is physically and economically possible is surprisingly close at hand.

You also said in your speech that our children "deserve better than the spectacle of censorship of the best scientific evidence about the truth of our situation and harassment of honest scientists who are trying to warn us about the looming catastrophe." There's also a second group of scientists involved in new energy research that has been suppressed even more and need to take their place in our quest for solutions.

New energy would shift the paradigm overnight. We will need public policies in place to:

1. Do the necessary R&D Apollo-style in secured laboratories, gathering teams of the best and brightest scientists and engineers in the field. Surprisingly, the investment in such an effort to produce viable practical prototypes would only be on the order of $200 million to $1 billion for five years, equivalent to 1-5 days of fighting in Iraq and a few days of profits for ExxonMobil. We must leave no stone unturned in this quest because the range of technologies is already broad and far-reaching.

2. Provide public forums to debate and discuss how to implement the most viable new energy options to mitigate climate change and pollution; and provide education and demonstrations for the public. We need to plan conversion scenarios that can help industry and government make the necessary transition to a new energy economy. The defense and aerospace conversion policies that I helped George McGovern, Fritz Mondale and Jesse Jackson draft during their campaigns, were minor compared to what we must do here.

While being politically incorrect at the moment, the consideration of new energy needs to be at the forefront of future energy policy discussions. It is too late to deny this, and we certainly don't want the control of these technologies be in the wrong hands by default. In President Eisenhower's words, "Only an alert and knowledgeable citizenry can compel the proper meshing of the huge industrial and military machinery of defense with our peaceful methods and goals so that security and liberty may prosper together." New energy needs to be controlled by We the People and so a strong grassroots movement will become necessary.

I cannot stress too strongly that an aggressive program to develop new energy is what humanity will need to survive our perilous situation. It may be painful for us to address these issues and may seem a bit far-fetched at first, but I can assure you these technologies are very real and can be developed as public policy. To that end, I have worked with Rep. Dennis Kucinich to draft legislation for providing public support for new energy R&D. Unfortunately, Congress is too distracted by other issues and too beholden to vested interests.

One final word: don't rely exclusively on those mainstream scientists, advisors, journalists and pundits who deny the reality of new energy. They are just as ignorant as those scientists who denied the practicality of aviation even after the Wright brothers were flying. But to expect the Wrights to immediately deliver a 737 would have been unrealistic.

In the conclusion of your speech, you said, "This is an opportunity for bipartisanship and transcendence, an opportunity to find our better selves and in rising to meet this challenge, create a better brighter future, a future worthy of the generations who come after us and who have a right to be able to depend on us." I couldn't agree with you more and we're on the same team.

Time is of the essence and we need to act soon. I look forward to your response.

Sincerely,

Brian O'Leary, Ph.D.
www.brianoleary.com

Chapter 13
Call for Innovators to Join the Solution Energy Revolution

"The resistance to a new idea increases as the square of its importance."
– Bertrand Russell

"Any sufficiently advanced technology is indistinguishable from magic."
– Sir Arthur C. Clarke

The world is at an energy crossroads. The alarming new information coming out of the climate science community confirms the unprecedented danger faced by all of humanity and nature by mankind's routine burning of hydrocarbons—oil, coal and natural gas. The resulting emissions of carbon dioxide and carcinogens into the Earth's atmosphere spell almost certain doom not only for the environment, but for human systems of government and commerce as we know them. Human survivability itself is in question, especially against the backdrop of vast deforestation, marine habitat destruction, accelerating species extinctions, and the threat from weapons of mass destruction on Earth, and, perhaps soon, in space.

Nature is fighting back with heat waves, super storms, rising and acidified oceans, desertification, species and disease vector migrations, and weakening of the Gulf Stream, in response to warming caused by injection of record amounts of carbon dioxide, methane and other greenhouse gases into the atmosphere. Despite this, and in the face of dwindling supplies of hydrocarbons, humans still consume as if there were no tomorrow. Even modest international agreements such as the Kyoto Protocols are ignored by the most polluting nations, especially the United States government, which seems to be more interested in going to war for oil than transforming its energy infrastructure to cleaner sources.

This multi-trillion dollar fossil fuel juggernaut, including going to war for oil, is the largest economic engine ever made in human history. We see record profits for the petroleum and war industries while innovation is stifled and largely ignored by established scientists, leadership and media. Yet innovation in our energy systems may be the single most important factor for our survival.

Significant solutions using conventional technology have proven to be elusive, prompting some scientists and environmentalists such as James Lovelock, Stewart Brand, John Holdren, Nathan Lewis, Richard Heinberg and myself to conclude that even the traditional renewables such as solar, wind, biofuels and hydrogen are not adequate to replace hydrocarbon combustion. Solar, wind, waves, tides, ocean-thermal, geothermal, hydropower and satellite solar power can suffer from intermittency, site unsuitability, diffuseness, limited availability and materials- and land-intensity. Biofuels such as ethanol and biodiesel compete with agriculture for land and still inject carbon dioxide into the atmosphere, albeit not as much as hydrocarbon combustion. Hydrogen is expensive to produce. It most often requires more energy to extract hydrogen than you get out of it, making this fuel an energy carrier but not an energy source. Typical methods of production (reformation of methane and electrolysis of water) still consume fossil fuels, emit carbon dioxide and can deplete atmospheric oxygen.

These fundamental physical limitations have led James Lovelock, Stewart Brand and others to reluctantly conclude that we should construct centralized nuclear power plants throughout the globe to produce electricity through grids in an electric economy. But because of limited supplies of uranium, high costs, hazardous fuel cycles and nuclear proliferation concerns, many of us in the scientific community (e.g., Union of Concerned Scientists, Bulletin of the Atomic Scientists, Federation of American Scientists) believe this is a very poor choice for our future. First, the questionable safety of nuclear power plants, especially in the age of terrorism, presents grave dangers to us all. The 1987 Chernobyl accident should provide us ample warning. Moreover, no safe long-term method has yet been found for disposing of high-level, long-lived radioactive waste—an inevitable byproduct of the nuclear fuel cycle. Finally, the proliferation of the technology throughout the world, would inevitably lead to acquisition of doomsday nuclear weapons by numerous irresponsible parties.

The prospects for "hot" nuclear fusion are equally dim. In spite of tens of billions of dollars over decades being spent on trying to achieve energy "breakeven" using gigantic Tokomak reactors, the results have thus far been negative. Moreover, nuclear fusion plants would constitute oversized, vulnerable facilities necessitating the continued use of ugly, antiquated centralized grid systems.

When full life-cycle environmental costs are considered, none of the above technologies appear to meet the criteria of sustainability— absent a breakthrough. By choosing any or some of them, we could only hope for incremental changes in our energy supply in the face of accelerating global demand. More importantly, these alternatives do not address the urgent time factor requirements for clean energy needed to mitigate global warming.

On the other hand, many new energy technologies have already been proven in hundreds of demonstrations in laboratories scattered throughout the world. Any one or some of these approaches, if properly developed, could end our dangerous dependence on hydrocarbons and uranium. Clearly, the traditional technologies keep us mired in the nineteenth and twentieth centuries rather than launching us forward into the twenty-first century. Nevertheless, this conventional thinking continues to dominate the news these days. Despite the great need, suppression of new energy has been historically documented in great detail by those who have taken the time to investigate. Inventors have suffered funding cuts, threats, sabotage and even assassination ever since the time of Nicola Tesla more than one century ago.

We define "new energy" to generally mean innovative technologies with the potential of providing a quantum leap in our ability to tap cheap, clean, safe and decentralized energy for producing fuels and electricity. These may or may not be recognized by mainstream science. The technologies include cold fusion, vacuum energy, advanced hydrogen chemistries, and energy from the thermal environment.

Any one of the above approaches to new energy promises orders of magnitude increase in cleanliness and affordability. In addition, there are many important transitional technologies that can mitigate emissions in the very near future: recycling and sequestration of carbon at the source through innovative chemistry and remediation of nuclear waste based on the principles of low temperature non-radioactive nuclear transmutations.

All of the above concepts (described in the last chapter) have already been demonstrated in laboratories throughout the world (I have seen many such demonstrations), and have been published in the peer-reviewed literature. But implementing them has proven difficult because there is no significant support. This lack of support for outside-the-box thinking is familiar to those of us who know the history of innovation. That is to say, there is generally a bias against the credibility of a new technology until it is accepted by the mainstream culture. The most strident objectors are often scientists themselves because some of their treasured "laws" appear to be broken by breakthrough experiments that often lead to profound technological change. And, as Russell stated in the quote at the beginning of this essay, the bigger the change the bigger still is the resistance, by a large margin. In spite of these severe limitations, I propose here that the transformation of our energy culture to one based on new energy is necessary for our survival, and that we should embark on a research and development program as soon as possible.

History is replete with examples of disbelief of new technologies when they first emerge. One example is aviation during its early days (Chapter 1). During those times, we had been embroiled in a vicious cycle of media and scientific blackouts of reality.

Unfortunately, the leading innovative nation, the United States, is living in fear since this century opened, with the inauguration of George Bush as its unelected president and its violent overreaction to the attacks of September 11, 2001. The nation appears to be too distracted by wars, repression, and the dominance by large corporations who don't embrace technological change outside of their own interests. The public awareness of the gravity of the global environmental crisis and the innovative spirit of America has gone underground, awaiting the opportunity to be sanctioned by the larger culture.

There is much discussion now about how the warnings we hear from leading atmospheric scientists continue to be ignored and scoffed at by those in power. In a refreshing counterpoint to politics-as-usual, former U.S. vice president Al Gore recently said that our children "deserve better than the spectacle of censorship of the best scientific evidence about the truth of our situation and harassment of honest scientists who are trying to warn us about the looming catastrophe." Yet there exists a second group of scientists involved in new energy research that has been suppressed even more. These truly unsung heroes of innovation will eventually take their place in our quest for solutions.

New energy would shift the paradigm overnight. As I outlined briefly in the previous chapter, we will need public policies in place to:

1. Do the necessary R&D Apollo-style in secured laboratories, gathering teams of the best and brightest scientists and engineers in the field. But first we should support a wide variety of inventors and technologies throughout the world. Surprisingly, this seed effort would only be on the order of $1 billion for the first few years, equivalent to a few days to weeks of fighting in Iraq or profits for ExxonMobil. Funds could come from public and/or private sources. At the moment, the new energy researchers receive no public support and only scattered private support. This is because of the fear element and that we are still on the toe of the profit curve and therefore in great need of public and/or angel funding. The seed money can come in the form of small business grants and loans to the 100-200 most promising researchers until they can attract capital or open source their technologies. As the technologies mature, we can expect the actual amount of investment and return to end up being significantly greater, depending on a number of factors other than the true R&D costs. The goal is to produce prototypes for the marketplace as soon as possible. Whatever management model emerges, we must leave no stone unturned in this quest because of the urgency of the global crisis. Fortunately, the range of technologies is already broad and far-reaching. The research effort should be international in scope and be immune to the political vicissitudes and corruptions of leadership and corporate dominance in the United States and elsewhere. Therefore, the research may need to be done discreetly at first under responsible and publicly accountable auspices. A governing body such as the United Nations should oversee the research, as no important resource like energy or water or food production or forest protection should be totally privatized.

2. Provide public forums to debate and discuss how to implement the most viable new energy options to reverse climate change and pollution; and provide education and demonstrations for the world community. We need to plan conversion scenarios that can help industries and governments make the necessary transition to a new energy economy, free of corruption and monopoly. We should assess the full life-cycle environmental impact of each alternative and its safety on a level playing field. We don't want to repeat the mistakes of touting its benefits without properly assessing its dangers and hidden costs (as in the case of nuclear power).

While being politically incorrect at the moment, the consideration of new energy needs to be at the forefront of future energy policy discussions. It is too late to deny this, and we certainly don't want the control of these technologies to fall into the wrong hands by default. In former U.S. president Dwight D. Eisenhower's words, "Only an alert and knowledgeable citizenry can compel the proper meshing of the huge industrial and military machinery of defense with our peaceful methods and goals so that security and liberty may prosper together." New energy needs to be controlled by the citizens of the world and so, in my opinion, a strong grassroots movement will become vitally important.

I cannot stress too strongly that an aggressive program to develop new energy is what humanity requires to survive this perilous situation. It may be painful for us to address these issues and may seem a bit far-fetched at first, but I can assure the interested reader that these technologies are very real and can be developed as public policy. To that end, some of us have been working with the U.S. Congress to introduce legislation which would support new energy R&D. Unfortunately, in these times, the entire U.S. government has been too distracted to work in the public interest.

We shouldn't rely exclusively on those mainstream scientists, journalists and pundits who deny the reality of new energy. Some of these skeptics do not seem to understand that we are in the research phase of an R&D cycle, and we cannot expect yet to have the kind of commercial prototype demonstration they desire in order to be convinced. They are just as ignorant as those scientists who denied the practicality of aviation even after the Wright brothers were flying. But to expect the Wrights to immediately deliver a finished product would have been unrealistic—or insane.

But, for the sake of argument, let us grant for a moment the remote possibility that the skeptics are right and that no new energy source were to prove to be practical for one reason or another. Would doing the research have proven to be a waste of time and money? Of course not. The path of discovery always comes up with unexpected surprises, and I would opt for such a modest effort, compared to the costs of war and polluting energy, when our survival is at stake. It is time to put altruism and creativity ahead of near-term profit.

Meanwhile, because of the urgency of the problem, I would encourage innovators throughout the world to move ahead to organize themselves to team up, obtain the necessary resources and perform research and development of new energy—in spite of cultural pressures to act otherwise. All of us should become educated about the possibilities and collectively support these pioneers of innovation, because we need all the help we can get to convert civilization from a catastrophic energy age to a new energy age.

This essay was first posted at the invitation of the World Innovation Foundation's web magazine *Scientific Discovery*. January 2007 issue.

Chapter 14
A Special Message for the Younger Generation about Solution Energy: Address to the United Nations Youth Assembly, August 2007

"We are not inheriting the Earth from our parents, we are *stealing* it from our children."

– David Brower

"Our current generation is committing *treason* against future genera-tions by destroying our global environment."

– Norman Cousins

"If…man encroached upon Gaia's functional power to such an extent that he disabled her, he would wake up one day to find he had the per-manent, lifelong job of planetary maintenance engineer…then at last we should be riding that strange contraption, the "spaceship Earth," and whatever tamed and domesticated biosphere remained would be our "life-support system."

– James Lovelock, *Gaia*

In this chapter, I depart even more from my usual professional restraint. I believe our dire situation calls for a much stronger expression.

Over the past 35 years I have researched both natural and break-through solutions for our sustainable survival. During the early years of my newly-found environmentalism, I became an advisor to presi-dential candidate Morris Udall and his U.S. House Subcommittee on Energy and the Environment. We exposed the dangers of nuclear power and the merits of renewable energy and increased energy efficiency.

I was somewhat older then than you are now. I was idealistic and took to heart these warnings of the late great environmental thinkers quoted above. They even inspired me to change careers at that time.

I'm sorry to say, things have only gotten much worse since then. I'm sorry to report that the state of the world you are inheriting is poised for utter destruction, largely because of the neglects of powerful people in my generation. Like a frog in a pot of water whose temperature is slightly raised every day until it's too late to get out, our home planet is reaching the boiling point. I'm sorry we handed you such a mess.

The good news is, the solutions are there, waiting in the wings to bring forward *now*. The bad news is, they have been suppressed by the most destructive empire in world history.

For example, I have traveled the world and visited some of the most promising demonstrations of "solution" energy technologies that could provide us with a quantum leap in our ability to have clean, cheap and decentralized energy in the near future (e.g., vacuum energy, cold fusion and advanced hydrogen and water chemistries). I have also discovered that many of these courageous inventors have been isolated, threatened and sometimes assassinated.

By "solution energy," what we really mean is making wise choices from among a wide range of nonpolluting options, free of vested interests. I believe that breakthrough, or new, energy is by far our most promising solution energy. As a default, we should also consider the cleanest possible blend of traditional renewables, such as solar, wind, tides, waves and hydrogen fuels provided from these sources.

As my research deepened, thus revealing what we must do to sustain ourselves physically, so too our social, academic, political and economic organizations have solidified themselves into a resistance so strong, we cannot seem to get off the train that is hurtling towards sure disaster.

When ordinary, moral and innovative people weren't looking, our beloved republic-turned-protofascist superpower has sunk into a morass of a criminal executive, a supine Congress, a corrupted judiciary, a bought-out media and an allowing public, collectively bent on totalitarian genocide and ecocide rather than embracing true solutions.

We cannot, *will not,* allow this to happen any longer. The perpetrators must be removed from power. And the whole structure of our decayed system must be revitalized in positive directions. The U.S.

Constitution and Bill of Rights and international charters, conventions and treaties must be restored and re-invigorated.

In the U.S. we need a revolution, a nonviolent one. Citizens will need to hit the streets in large numbers demanding change. A new American Revolution has been long in coming, with warning signs coming from so many presidents and others for more than two centuries of a gathering storm that is reaching a climax now.

But how can we have a revolution if so few of us show up for it? On the fifth anniversary of 9/11, I spoke at a peace-and-truth rally in front of the White House (Chapter 20). Barely thirty nonviolent demonstrators marched and gathered around the podium, while two snipers prowled across the White House roof. It was a nice day to come out, yet only one American in ten million joined our motley crew. Why so few? Do we care?

If this were Latin America, millions of people by now would be jamming Washington and banging pots on a constant vigil to demand the resignations of the president and his peers—until they leave. But not in the United States we have now. Whatever has happened to the kind of activism we have seen in the courageous leadership of a Patrick Henry or a Martin Luther King? In the absence of correct government behavior, the people have no choice but to take to the streets and stay there until we have the needed results.

Democracy is not a spectator sport. It can only die if we sit on the sidelines.

I am also astounded that President Eisenhower's 1961 warning about the severe consequences of the buildup of a military-industrial complex has gone unheeded. Now, almost fifty years later, this awesome power has become much further institutionalized and centralized into a hydra-headed beast opposed to all life on Earth—except for their own lives. Only we the people can remove this power and take control of our collective destiny.

But the crisis also presents opportunities we may have never imagined. First, though, we must clear the way by once again declaring our independence from King George and his cronies in power. We must impeach, remove, and convict them. We must throw their tea overboard if they don't slink away peacefully.

The reckless actions of the Bush administration are sealing the fate of an entire planet— unless we have the courage to stand up and say "**NO**". It's our time to re-create our republic, it's our time to invite the world to join us in a new republic whose goals are life, liberty and the pursuit of happiness. Basically, we deserve to create for ourselves a peaceful, sustainable, just future and nothing less.

Our top priority is to demand the resignations of BushCo, or that Congress should impeach the wrongdoers. Now. This is an essential first step towards restoring Constitutional government. All of us here should insist that impeachment be put back "on the table." Congress has not yet put it "on the table" because of their fears that the table which also supports their own careers might collapse from under them. What a rickety, corrupt "table" this is.

The only items now "on the table" are those that our dictators have piled up for us, a bill *your* generation will have to pay. Within this young century alone, that bill is greater than a trillion dollars more of weapons on Earth and planned for space, a trillion dollars more for illegal, immoral and preemptive wars in distant lands, a trillion dollars more for oil profits, a trillion dollars more in the pockets of cronies and speculators, a trillion dollars more of national debt, a trillion dollars more for the anti-Muslim lobbies, a trillion dollars more for the legal and illegal drug trade, and so on.

The new table we must take out of the closet is really a hearty old table that needs dusting off. It is the U.S. Constitution and Bill of Rights, which have helped make our republic survive for this long. The Constitution is endangered by a tyranny so great, it is hard to imagine the consequences: the wanton environmental destruction, the unrestrained torture, the suspension of habeas corpus, the unprovoked wars of conquest, the resource exploitation in the global South: and the fixed elections, the potential for martial law and a police state dictatorship at home. Meanwhile, we only seem to hear imperial lies, lies and more lies, dutifully reported by a propagandistic media.

Our time calls for a number of systemic actions even beyond a cleaning out of those who have betrayed their oaths of office. It is a call for a restoration of proper vision.

Al Gore has recently given those of us in the industrialized world a mandate that he and many leading climate scientists feel will be necessary for our survival: act radically within ten years, leading to a reduction of carbon dioxide emissions of more than 90% by 2050 . As an atmospheric scientist myself, I agree with that mandate, and I also believe that the solutions will come from outside the box of our retrogressive thinking.

But we need more: a restoration of truth in our public discourse. For only the truth can set us free from the tyranny that so imprisons us. We need to look squarely again at 9/11 truth, at depleted uranium truth, at war pretext truth, at torture truth, at electoral fraud truth, at solution energy truth, at UFO/ET truth, and all the other truths that our cancerous tyranny is trying to destroy.

And we will need to reconcile all these truths, which are as self-evident as they were when we declared our independence 232 years ago. We must eventually reconcile ourselves with ourselves and with the natural world. We Americans must apologize to the rest of the world for our aggressions and neglects.

As an elder American citizen, I owe you two apologies, generational and global, for the transgressions committed by some of my peers.

We can set goals, cajole governments and create NGOs geared towards peace, sustainability and justice. But we can never get close to achieving those goals by these modest approaches alone. The imperial system with its war machine can overwhelm the efforts of even the most well-intentioned individuals and groups.

Our activism needs to go beyond our own parochial issues, which are based more on personal recognition, careers, advancement and the temptations of money. Albert Einstein said, "Try to become not a man of success, but try rather to become a man of value." This becomes ever evident as we organize ourselves and move into our higher collective purpose.

Our movements have become anemic, divided and conquered. We will need to ally our movements and declare those values that cut across all ethnic and interest groups, values that ring so true we cannot help but come on board. Values such as peace, sustainability, justice, freedom and equality for all.

But to achieve all that, we must go beyond declaring our independence from King George. We must declare our *interdependence* with each other and with all of nature in the global village. We must declare ourselves a sovereign world community comprised of an alliance of movements of various ethnic and religious identities who have a common and abiding interest in survival and good health. We must force our governments serve us, not the other way around.

Beyond our apologies, we Americans also have an opportunity as never before to transform our imperialism into an invitation to found a new global green republic that embraces the higher purpose we all so yearn for but have been so cut off from.

To help facilitate the restoration of world peace and sustainability, Meredith and I have moved to the Andes of Ecuador and have built a retreat and nature center, Montesueños (Mountain Dreams). Here we have found fertile soil to co-create with nature and kindred spirits a prototype community for what-could-be locally and globally, far from the fears and distractions of a world gone mad. Our goals are world peace, sustainability and justice, our mission is research, education and action. Our values are truth, beauty and compassion.

We have found harmony here in Latin America as a crucible for new ideas, clear vision and radical social change. For example, we are working with the government of Ecuador to save the one-million-hectare Yasuni National Park, one of the most biodiverse rainforests on Earth, housing voluntarily isolated indigenous tribes and a superb botanical carbon dioxide sink. It also sits over a billion barrels of heavy crude oil, whose $30 billion extraction would ruin this precious Amazonian ecosystem. The thirst for the oil is overwhelming even though this resource would provide a ravenous world with only 12 days of supply!

No one issue compels our desire to embrace these new structures more than the possibility of a revolution in our energy policy. After twenty years and a lifetime of study, I am convinced everyone in the world can have clean, cheap, safe, decentralized energy, water, food and a healthy environment. I am also convinced we can preserve our forests, wetlands, oceans, biodiversity, cultural identities, food and water supply.

But we cannot do so under the boot of a voracious corporatocracy bent on destroying our precious habitat for their own selfish gain. I don't think I need to present the obvious statistics to prove my point.

It is essential to our survival that we openly examine the implications of having breakthrough new energy as one mega-solution to our crisis. We should create a forum for discussing this very real possibility, free from the shackles of suppression by those in control, unwittingly allied with an irrational scientific censorship. We must get beyond their debunking us as "pseudo-scientists" and "conspiracy theorists." The onus is on *them,* not on us, to try to demonstrate a negative. We're free to proceed, bearing in mind that the coveted "laws" of science are really only theories.

Let's also go beyond the focus on whether this or that gizmo will be the magic bullet or who will become the Bill Gates of new energy. Let's think beyond looking just at the technologies themselves. Let's go for the larger context: what do these developments mean for our collective future? What kind of social system can facilitate this?

Clearly, we cannot allow Cheney *et al* run this one too. When massive profit and exploitation are not in sight, they always have suppressed, and always will, suppress it as long as they remain in power. It is clear they are destroying the world with an energy policy that will go into infamy as the most biocidal catastrophe in human history—with the possible exception of nuclear war (including the ones now underway that use depleted uranium as a weapon; this is a disastrous cancer-inducing radioactive substance that will affect untold future generations, up to 4.5 billion years!). Such wars and energy policies must be stopped.

We must look towards a future that cannot or will not resemble the past if we are to heed Al Gore's requirement that we drastically reduce carbon emissions, if we are to reduce the perilous dangers of a nuclear energy economy, or if we are to lessen the extensive environmental impact of a biofuel economy that still emits CO_2 and takes away from our forests and agriculture. Even if we create a solartopia or wind economy, these require massive infrastructure, and use a lot of valuable land and raw materials. Any sensible energy policy demands the broadest possible look at *all* alternatives, free of vested interests.

But the possibility of "solution energy" demands that we restore our democratic systems worldwide. It mandates social structures that can facilitate and regulate these technologies—if they are indeed found to be sufficiently clean and cheap—so they can become available to all of humanity. The main obstacle to a Solution Energy Age is a human question, not a technical one.

In summary, an energy solution revolution requires first a revolution overturning the actors and policies of our powerful and suppressive governments and corporations.

Short of divine intervention, we have no other choice but to act, if we are to survive this crisis. This goes beyond morality. It is necessary logic.

Humanity has moved nature to a tipping point, maybe beyond, according to leading atmospheric scientists and ecologists. Our current focus on fear, economic turmoil and wars of conquest can only deepen the crisis. How we solve these problems demand of all of us the most revolutionary thinking and action we have ever dreamt of.

So let's jump ahead to the year 2050. What kind of world will you be inheriting then? I can see two extreme possibilities: the one where we are now headed and the one in which we (you) take the reins of power and declare a solution energy future   The first is a totalitarian world in which the last drops of oil are squeezed out of the rainforests of the Amazon, the tundra and waters of the Arctic, and the tar sands of Alberta. A world of mass migrations from inundated coastal cities, a world without icecaps and glaciers, a world in which jungles become deserts, a world with few fish, a world where, ironically, Europe is placed into a deep freeze by the shutting down of the Gulf Stream conveyor system. A world devoid of fresh drinking water except for the rich. A world of cancer-inducing radioactive residue, toxic air, water and soil.

Or you could have a global green democracy and/or a revitalized United Nations. A world of clean, cheap energy, food, water, and quality healthcare and education for a slowly dwindling population, now under benign control. A world in which you carry the torch as the new leaders and honest politicians of the future. You are the heirs of a reborn Gaia. The air and water are clean now. Carbon emissions are reduced by over 90%. People are happy and free. Nature is preserved, restored and sustained.

If you're 24 years old now, you will be my age (67) in 2050. I'll probably be long gone. The choices and consequences are yours to make! As a scientist and concerned world citizen, I have presented many things we must do to avert global catastrophe. How we implement our plan will be up to all of us, and will take great vision and courage.

We should start a "conspiracy of youth" to do the necessary long-term planning to create a peaceful, just and sustainable planet. Remember, the literal meaning of conspiracy is "breathing together," something we know how to do in the practice of yoga. This kind of approach will become necessary if we are to survive.

We have a very big job to do, so let's roll up our sleeves and get going...and Godspeed!

This chapter is based on a speech given to the United Nations Youth Assembly, August 15, 2007.

Chapter 15
Conventional Alternative Energy Promotions — None of the Above:

How authorities conceal breakthroughs that can help us survive the ecological crisis

"The Kyoto Protocol totally avoided the material challenge of stopping activities that lead to higher emissions and the political challenge of regulation of the polluters and making the polluters pay in accordance with the principles adopted at the Earth Summit in Rio. Instead, Kyoto put in place the mechanism of emissions trading which in effect rewarded the polluters by assigning them rights to the atmosphere and trading these rights to pollute."

– Vedana Shiva, www.zmag.org/znet,
Dec. 13, 2007

"Like biofuels and micro wind turbines, carbon capture and storage turns out to be another great green scam. It will come too late to prevent runaway climate change; the government has no intention of enforcing it; and even if it had, the technique is likely to boost our carbon emissions."

– George Monbiot, *The Guardian,*
March 18, 2008

"If the entire national corn crop were used to make ethanol, it would replace a mere 7 percent of U.S. oil consumption...more than 40 percent of the energy contained in one gallon of corn ethanol is expended to produce it (and comes) mostly from oil and natural gas...the environmental impacts of corn ethanol production are serious and diverse. These include severe soil erosion of valuable food cropland, plus the heavy use of nitrogen fertilizers and pesticides that pollute rivers. Fermenting corn to make one gallon of ethanol produces 12 gallons of noxious sewage effluent...The science is clear: The use of corn and other biofuels to solve our energy problem is an ethically, economically and environmentally unworkable sham."

– Prof. David Pimentel, Cornell University,
"Corn Can't Save Us," *St. Louis Post Dispatch*,
March 18, 2008

"Turning tar sands into oil requires almost as much energy input as they contain at the end of processing...the process begins with clearcutting the boreal forest, destroying habitat and soil...Tar sands production threatens to turn much of central Canada's water reserves into oily wastes unfit for consumption...the rush to mine tar sands (www.oilempire.us) resembles an indigent cigarette addict looking through ashtrays to find a couple of butts that can be relit to get a couple final (nasty tasting) drags of tobacco smoke."

– Mark Robinowitz,
www.globalresarch.ca,
Dec. 11, 2007

"The dirty secret is that nuclear power makes a substantial contribution to global warming. Nuclear power is actually a chain of highly-intensive industrial processes. These include uranium mining, conversion, enrichment and fabrication of nuclear fuel; construction and deconstruction of the massive nuclear facility structures; and the disposition of high-level nuclear waste (Michael Lee, Council on Intelligent Energy & Conversion policy)."

– Karl Grossman, "Money is the Real Green Power:
The Hoax of Eco-friendly Nuclear Energy," *Extra!*,
Feb. 3, 08

"Silicon (for solar power cells which produce 0.1% of the world's energy) is in short supply."

– Tatsuo Saga, Sharp Corporation, 2008

"The rhythmic whoosh, whoosh, whoosh of wind turbines echoes through the air. Sleek and white, their long propeller blades rotate in formation, like some otherworldly dance of spindly-armed aliens swaying across the land...For many, the realities of living with wind turbines are more complicated than (claims of) clean energy and easy money."

– CNN, Aug. 19, 2008

"At the World Renewable Energy Conference in Glasgow I recently witnessed the strange phenomenon of group denial first hand. After a paper about hydrogen-fueled cars, some embarrassing questions were asked about the practicalities of storing and delivering hydrogen to the cars. The questions were dismissed and the questioners meekly backed down. I wanted to jump in and set them straight but keenly felt the group pressure to not ruin the party. I couldn't do it! Groupthink is a strange phenomenon resulting from our deep genetic programming as herd animals: If our peer group is ignoring the giant lump in the living room rug, we will naturally imitate their behavior and walk around the elephant hidden there. We tend to be drawn into a sort of mass hallucination where everyone conforms to an unspoken agreement to ignore the inconvenient but obvious truth. We walk around the lump without consciously seeing it."

– Thomas R. Blakeslee, "The Elephant Under the Rug: Denial
and Failed Energy Projects," www.RenewableEnergyWorld.com,
Sept. 2008

N uclear power. Carbon sequestration at coal plants. Ethanol-from-corn. Other kinds of biofuels. Carbon cap-and-trading. Hybrid cars. Conventional electric cars. Air cars. Gas-turbine micropower. Efficient powerplants. Hydrogen economy. Hydro-power. Geothermal energy. Solar. Wind. Tides. Waves. Ocean thermal gradients.

Which one(s) of these will solve the climate crisis and give us a large and lasting contribution to energy sustainability? The sobering answer to any truthful inquiry, I am sorry to say, is none of the above. As I mentioned earlier, we, in our mainstream discourse, are almost literally tilting at windmills.

On a daily basis, I've been asked to vet this or that new macro-scale "renewable" technology that, upon further examination, doesn't turn out to be very renewable after all, especially when it comes to replacing the world's current massively dirty energy policies. Even in small doses, none these concepts could make an appreciable dent into what must happen in our future energy mix. Almost all ideas originate from a kind of groupthink that comes from a commercial promotional bias that says, "maybe this will work, or maybe that will work." Rarely are the drawbacks considered.

Nevertheless, we see some possibilities that can make marginal differences. For example, thin film solar photovoltaic technology now being researched stands head and shoulders above silicon cells. Biodiesel from algae makes much more sense than ethanol from corn and other crops (e.g., http://www.valcent.net/i/misc/Vertigro/index.html).

But even here, we see industrial promotions and no dispassionate analysis and vetting. For example, biofuels from algae also suffer from requiring a large infrastructure, extensive land-and-water use, and the problems of contamination.

Lester Brown's "Plan B" for future renewable energy policy suffers from the same marginal effects. He too goes for partial macro-solutions, such as wind-powered electric cars. (http://www.earth-policy.org/Updates/2008/Update75.htm)

Brown's advocacy of these incremental possibilities at least has the value of pointing out the deep fallacies of current choices. Brown's and his colleagues' critiques that mainstream alternative projects such as nuclear power, "clean coal," biofuels and hydrogen cars are not the best ways to go do "make the news" in environmental and progressive circles and put the vested powers on notice that their own touted and well-funded "alternatives" are themselves dirty and expensive boondoggles. Considering scenarios such as Plan B help point the way and provide an improved frame of reference for evaluating what we really need to do. These are the more sensible categories of well thought-out "first generation renewable energy" sources, going for the best of today's understood and accepted technologies.

What we need is a Plan C: only second-generation renewable energy such as zero-point, cold fusion and advanced hydrogen and water chemistries, have the potential to reverse humanity's perilous course towards runaway climate change and irreversible pollution. Yet nobody with significant public access knows or wants to know this. Why is there such a disconnect about this? During a reception at an alternative energy conference we attended in 2000, why did Brown and his colleagues turn away from me when I gently broached to them the concepts of breakthrough energy?

In my 69 years in this embodiment, I have experienced increasing awareness from within as well as decreasing integrity among policy-makers from without, about how humans have lost the capacity to cope with the challenges of our war on humanity and nature. My native country, the United States, leads the way into this insane destruction of ourselves and our habitat.

The most conspicuous two examples now are our inability to extract ourselves from genocide in the Middle East and to answer the clarion call to do *something* about global warming, climate change and carbon emissions. Those in the public eye stop short of the solutions we so desperately need to reverse the climate crisis, even among the so-called environmentalists themselves.

Moving into lasting, elegant breakthrough energy solutions is denied even by those who should know better—the mainstream progressives, environmentalists, scientists and investigative journalists. They, too, seem to be defending their own vested interests in incremental remedies, which, like the Kyoto Protocols themselves, would lead to results that are too little and too late.

The deception, whether conscious or not, is so glaring, the public has become bewildered by fear, disinformation, obfuscation and conformity to politically correct but environmentally catastrophic norms. By default and through clever propaganda, we are blinded from waging pre-emptive war on what it will really take to solve the climate crisis, because nobody of influence wants to bite the bullet of the profound changes we need to make. Meanwhile, our corporate masters accumulate ever more wealth and the power to suppress innovation as we go down the road of denial and destruction.

I sometimes liken this process of problem-solving to peeling an onion whose successive layers contain increasing germs of truth and decreasing levels of conformity. An unpleasant process of tearful revelation emerges only for those who choose to look. Most of us don't want to navigate beyond our own comfort zones out of fear of personal consequences, ranging from the loss of friendship and career to the prospect of humiliation, imprisonment, and even torture or murder.

So, by default, most of us give into living the lie of a false political "consensus" that represents only the outermost layers of the onion, sealed by the tyranny of the elite.

I call this model "concentric circles of conformity versus truth". We have been led to believe that we are in a state of perpetual war in an Orwellian world in which our leaders lie their way through power. And, as we all know, truth is the first casualty of war. Conformity rather than truth becomes the standard of our behavior. We dumb down.

The outermost layer contains the most blatant falsehoods, controlled by a corporatocracy of money, weapons, dirty energy and "security." The Republican neocons of the Bush-Cheney regime and their financial supporters have effectively carried this banner. Judging from the most lucratively funded, media-dubbed Democrats in Congress and presidential "front-runners," they, too, are waltzing into their own Tweedle-Dum roles as the new perpetrators of the lies of war, tyranny, pollution and fiscal irresponsibility.

But the next layer into the onion, which I call the standard progressive critique of the way-things-are, can be even more deceptive. Going after the politicians and corporations in charge, these individuals and groups, whose expression can be eloquent, nevertheless offer no significant or lasting remedies to our massive carbon dioxide discharges into the atmosphere.

Yes, the progressives might even give up on the Democrats and vaguely seek political strategies. They want the war to end. They want the illegal surveillance, secrecy and torture to stop. Some of them want to impeach the president and vice president. But, based on results, they too are spineless and so the noises they make are just as irrelevant to what we need to do.

The progressives and mainstream environmentalists seek 1970s incremental energy solutions such as solar, wind and increasing efficiency without having the foggiest idea that breakthrough energy such as zero point, cold fusion and advanced hydrogen chemistries wait in the wings with *real* answers that could give us clean, cheap and decentralized energy.

In 1975, I served as energy advisor to presidential candidate Morris Udall. I also worked with some of the Democratic majority on his U.S. House Subcommittee on Energy and the Environment, where we envisioned a largely renewable energy economy by 2000. We felt we could accomplish this with a dramatic increase of solar, wind and geothermal energy, plus improvements in efficiency. These actions would allow us to replace the burning of hydrocarbons and uranium with much cleaner energy.

The Congressmen and I felt these approaches could some day become the mainstay of energy use, using today's technologies bound by the traditional "laws" of physics. But such a Manhattan-type commitment would require new investments on the order of trillions of dollars reallocated from defense spending and dirty, unsafe energy that dominated, and still dominates, our energy policies. Going for the renewables would be a hard sell (especially in these times of economic pressures to the contrary), but feasible in the event we had a re-awakened Congress and a visionary administration.

But now, 33 years later, the results have been just the opposite. The vested interests in coal, oil and nuclear energy are soaring in profit and power. Solar, wind, and a renewables-generated hydrogen economy, while on the increase, still represent only about 0.2% of total global energy use (see the Appendix for details). Meanwhile, during that interval, we have *tripled* the burning of coal, oil and gas— this in spite of the warnings of the climate scientists.

And, now, we're finding out that solar and wind are not as renewable as they first appeared to be. Any careful study free of advocacy, shows these sources are too intermittent, diffuse, and materials-, land- and capital-intensive to come even close to meeting world energy demand. Geothermal and hydropower are also approaching their limits.

The solar and wind advocates comprise the bulk of mainstream clean energy thinking. My colleague Keith Lampe calls these solutions "first-generation alternative energy" or "transition energy." Lampe writes that the No-War-No-Warming activist element talks about these solutions "without mentioning-even briefly!-that none of these are adequate and that we must proceed with utmost pace to applications of second-generation renewable modes (new, or breakthrough energy technologies)."

Some progressive scientists and environmentalists criticize nuclear power, ethanol-from-corn, carbon cap-and-trading and carbon sequestration at coal plants. But they stop short of acknowledging the *possibility* of breakthrough energy. They are satisfied with a solartopia of "renewable" solutions, none of which are adequate to turn the tide. But why settle for less, when we have better choices? Why are we so paralyzed by such a paradigm?

From forty years' experience in analyzing these tried-and-true energy options, I join many of my scientific and journalistic colleagues like Nathan Lewis, John Holdren, Richard Heinberg, George Monbiot and many others to assert that the bulk of our future energy can come from **"None of the above."**

By that, I mean that renewable energy is not really renewable when full life-cycle environmental costs are considered. How could ethanol-from-corn be taken seriously when scientists such as David Pimentel of Cornell calculate that even more carbon emissions come from burning our food than from burning hydrocarbons (when we include planting, harvesting, transporting and storing the fuel) and can only exacerbate hunger and high food prices? The fact that one SUV tankful of ethanol could feed one hungry person for one year should make any of us with any compassion to stop there.

Nuclear power is an unmitigated disaster. It is expensive, dangerous and has the potential for creating monstrous weapons that could end civilization. We don't even know how to dispose of radioactive waste that could be with us for tens of thousands of years. A meltdown from a nuclear power plant "accident" or attack could spread lethal radiation to millions of people and cost the public on the order of $1 trillion.

Carbon sequestration at coal plants is a joke. The Bush Administration wants to spend $324 billion on the coal industry to blast their dirty emissions back into the ground in a macro-engineered terror-forming grossness, reminiscent of removing mountaintops, nuking and zapping the ionosphere with radio waves, or spewing particles into the atmosphere to try to offset global warming. Even if these gross-level carbon sequestration technologies were to work, they wouldn't come on-line until 2030. According to NASA climate scientist James Hansen, this would already be too late to reverse climate change or to prevent the melting of the Greenland and Antarctica ice caps, resulting in tens of meters of sea level rise and the inundation of the world's coastal populations.

My answer to all the Ponzi schemes and industrial handouts—and to even the solartopia solutions, in the long run, is **None of the above.** Cruising through the layers of discourse all the way between the corporatist Republicans and Democrats to the progressive critics and establishment environmentalists, can only seem to uncover, at best, a set of transitional strategies which could increase from a few tenths of one percent now to a few percent later, of the world's total energy mix (Appendix).

We must dig into deeper layers for our answers. They lie beyond the realm of "respectable" political discourse, for we are dealing with physical problems that demand physical solutions which no amount of hubris or propaganda could dispel. Undeveloped, unsupported breakthrough solution energy technologies are everywhere waiting for their opportunity to avert the climate crisis and rampant air pollution. But King Cong industrialists (coal, oil, nuclear, gas) don't want us to know that. So, with all the zeal of the powers-that-be, they systematically suppress the new technologies through threats, assassinations, ridicule and no support—and they provide lip service to half-measures such as biofuels, tar sands, carbon cap-and-trading, wind and solar.

I believe the next layer beneath the standard progressive critique and solartopia proposals contains the key to understanding what we must do. This layer represents uncovering truths that are even more inconvenient than Al Gore's call for a 90% reduction in carbon emissions by 2050. These unsung layers include new energy truth, water resource truth, biofuels truth, 9/11 truth, war pretext truth, depleted uranium truth, electoral fraud truth, and a host of other truths awaiting our analysis, if we only have the courage to step outside the box of conventional thinking and political correctness.

Yet my repeated attempts to call attention to the *possibility* of breakthrough energy fall on deaf ears, ranging from Ralph Nader's Critical Mass energy policy group, to leading progressive Democrats—even Al Gore himself, so far. In each instance, I have gotten a courteous dismissal or no answer at all. To entertain this possibility is to be perceived as a scientific outcast or conspiracy theorist. This is career-destroying stuff, so those energy scientists charged with having the best new ideas really don't have the best new ideas, and so join the established elite in unwitting alliance as guardians to the gates of acceptable public discourse and scientific credibility.

We are now caught in a stalemate of rhetoric without action. Those of us operating outside-the-box have been suppressed. The most important truths lie deeply within layers of deception and distraction while business-as-usual prevails and nature dies.

Yet the deeper layers *must* be uncovered if we have even a prayer to survive. Do we have the courage to do that, to question authority and to seek those answers that lie outside this suffocating conformity? In my opinion, we have no other choice but to address these deeper issues and to act soon.

Some of this chapter is based on an essay posted on November, 2007

Chapter 16
What Kinds of Energy Sources
Do We Really Want?
Let the Dialogue Begin

Our choice of future energy sources should be openly debated. At the very least, they should be reliable, cheap, clean, safe, decentralized and publicly transparent. We must transform our institutions to facilitate the transition of energy technology into the 21st Century rather than back to the 19th and 20th Centuries. The old ways won't work anymore.

"The Bali conference featured an alternative movement-building component outside: a Climate Justice Now! Coalition...(They) criticized carbon trading and called for genuine solutions: 'reduced consumption; huge financial transfers from North to South based on historical responsibility and ecological debt for adaptation and mitigation costs paid for by redirecting military budgets, innovative taxes and debt cancellation; leaving fossil fuels in the ground and investing in appropriate energy-efficiency and safe, clean and community-led renewable energy; rights-based resource conservation that enforces Indigenous land rights and promotes peoples' sovereignty over energy, forests, land and water; and sustainable family farming and peoples' food sovereignty.'"

– Patrick Bond, www.Znet.org,
Jan. 6, 2008

"Our world, our old world that we have inhabited for the last 12,000 years, has ended, even if no newspaper in North America or Europe has yet printed its scientific obituary. ...What if growing environmental and social turbulence, instead of galvanizing heroic innovation and international cooperation, simply drive elite publics into even more frenzied attempts to wall themselves off from the rest of humanity? Global mitigation, in this unexplored but not improbable scenario, would be tacitly abandoned (as, to some extent, it already has been) in favor of accelerated investment in selective adaptation for Earth's first-class passengers. We're talking here of the prospect of creating green and gated oases of permanent affluence on an otherwise stricken planet...the current ruthless competition between energy and food markets, amplified by international speculation in commodities and agricultural land, is only a modest portent of the chaos that could soon grow exponentially from the convergence of resource depletion, intractable inequality, and climate change. The real danger is that human solidarity itself, like a West Antarctic ice shelf, will suddenly fracture and shatter into a thousand shards."

– Mike Davis, www.tomdispatch.com,
August 11, 2008

"Big business says addressing climate change 'rates very low on agenda.' (A) poll of 500 major firms reveals that only one in 10 regard global warming as a priority."

– T. H. Davis, G. Lean and S. Mesure,
www.independent.co.uk, Jan. 27, 2008

"Global warming is the privatization of global commons by capital which now involves the expropriation of ecological spaces of the South. Progressive climate strategy must reduce growth and (unsustainable) energy use while raising the quality of life of the broad masses of the people... the very existence of humanity and the planet depend on the institutionalization of economic systems based not on feudal rent extraction or capital accumulation or class exploitation, but on justice and equality."

– Walden Bello, www.znet.org,
April 8, 2008

"It is far better to devote our talents and investment dollars on hastening the arrival of its successor, than prolonging the agony of oil's decline...From now on, America's top priority in the energy field must be to explore all potential components of this new energy future and move swiftly to develop those with the greatest promise."

– Michael T. Klare, *Foreign Policy in Focus,*
June 28, 2008

"The combined total (cost of the Iraq war is) well over $2 trillion. Compare this with the $1.5 billion earmarked by the administration in 2007 for climate-friendly fuels…If we had any sense at all, we would terminate the war as rapidly as possible, reject all war-related supplemental funding requests, dramatically cut our reliance on petroleum, and transfer massive funds from Iraq war accounts to research on alternative energy systems."

– Michael T. Klare, *Foreign Policy in Focus,*
Dec. 10, 2007

As a former energy policy advisor and professor involved in energy issues for the past three decades, I have reluctantly come to the conclusion that no existing energy technology held credible by the mainstream can offer a solution that is even close to satisfying the criteria we need to meet the global demand for energy that is compatible with our survival. Only innovation can provide the answer. Even the $1.5 billion allocated to alternative fuels is mostly wasted on a biofuel boondoggle.

We the people must therefore take responsibility for what kinds of energy sources we'd like and how much of each we want to have in the future. We cannot continue indefinitely with hydrocarbon fuels and nuclear energy. And we must consider all ramifications of the full range of alternatives, known and unknown, of which the popular culture is still mostly ignorant. We have some important choices to make and we don't want them to be imposed on us out of the self-interest of elite individuals and organizations. We need to become educated about the issues, about all the possibilities that await us if we only give them a careful and unvested look.

The time has come for an intelligent dialogue to ensure the timely deployment of those energy sources that meet commonly agreed-upon criteria for the global public good. In the language of engineering design, these are called "requirements," which is the process of setting criteria for what we in the public commons would like to have and when. Like ordering food from a menu, we could order the mix with the best prospects for a clean and sustainable energy future.

Let us look at the criteria for a sane energy future in more detail. These openly-discussed technology assessments should rate the most promising energy systems within each criterion. Basically, we want to make new energy systems that are:

Feasible. Based on the available research data, we have many more options than is commonly understood, especially in the new energy technologies. We need all the engineering support possible to transform the early experiments into reliable and dependable reality. Supporting the R&D from altruism and public awareness will be important to the process. We can focus on technologies that could give us simplicity, operational reliability, timely energy delivery, long lifetime of use, and potential for scaling up or down.

Cheap. What are the capital and operational costs of the device? How soon and for how much investment can it be manufactured in the billions of units for global use? Is this ecologically supportable? How can we economically transform ourselves to a new culture of inexpensive energy while creating new jobs to clean the planet? For example, would it not be obviously wise to provide jobs in a renewed human-and-nature-centered infrastructure and in an Earth Corps to clean up the environment?

Clean. This is probably the most important consideration of all: What are the full life-cycle environmental costs of a given system? Operationally, some of us believe we can develop sources that can deliver close to zero emissions globally by 2020 if we do the necessary R&D very soon. Just as important are other environmental impacts such as, how energy-and materials-intensive is the source? What pollution would come from the extraction, manufacturing, transport, operations and safe disposal of the hardware? Is there waste material generated? How much manufacturing or growing is needed? How much land is used? What are the aesthetic implications? Do we need more power grids, power plants, refineries, solar farms, windmills, hydroelectric, biomass and hydrogen generators and storage facilities? A very cogent animation by Annie Leonard's popular "The Story of Stuff" lucidly describes the vast hidden costs of overusing our raw materials from cradle to grave: www.storyofstuff.com. We will need a detailed study of these three criteria of feasibility, low cost and environmental impact in our strategy for developing a sensible energy future.

Safe. We must find ways to avoid any weapons use or overuse of breakthrough energy. In an age of imperial and corporate hubris, terrorism and polarization, we do not want to see this technology be abused as in the nuclear story. So this means designing fail-safe systems that could deliver only so much power. Many environmentalists justifiably fear even bigger, noisier appliances, power saws, personal helicopters, etc. that would pollute the environment ever more. This development requires from the outset public transparency, acceptability and regulation of new energy technologies (see below). Is all this possible? We won't know unless we try.

Decentralized. Some of us believe that future energy sources can be localized by a selection and blend of low and high technologies. There is no reason why, in the long run, electricity must be brought in through antiquated grid wires or generated through large central station power plants, whether they were fuelled by hydrocarbons or carbohydrates or hydrogen or uranium or hydroelectric or any other source. There is also no reason to have to believe we need to rely on fossil fuels—and later on, biofuels and hydrogen, at depots for future transportation and heating and cooling of buildings, to the exclusion of direct new energy usage. We can be smarter than the conventional wisdom dictates in producing sensibly decentralized new energy infrastructures.

Publicly Transparent. In the absence of public awareness, oversight and regulation of global energy policies, we will surely continue to remain far off the path towards a sustainable future. This might be hard for vested interests to accept, but will become apparent to those of us with no prejudice as to which technologies will best serve the public interest and when. We can rise above this pettiness of policy once we realize that the unwise choices of the past and present have been dictated by elite special interests. To alleviate this disconnection, those of us unbeholden to the energy lobby will need to be the ones to objectively assess the technologies, educate, and create public dialogue as we confront the choices ahead (e.g., www.newenergycongress.org). We should create scenarios for the orderly transition from a polluting to a new energy economy.

This is only the beginning of a dialogue about selecting from the menu of options addressing the burning question, What do I want in a future energy policy? A feast of possibilities awaits us. Do we have the courage to explore, to choose, to overcome tyranny, and to enjoy the fruits of the effort? In this journey we need to be intelligent and wise in selecting the best course for us and for Mother Earth.

So what kinds of energy sources do we really want? As we shall see, the "we" need to be transformed to the "you." Most of the "we" are asleep in deference to a frightening tyranny that has shut us down. It's up to you. What kinds of energy sources do *you* want?

Most of this chapter is based on an essay posted in May 2004.

Chapter 17
The Hidden New Energy Debate:
Will We Want to Use It, Abuse It,
Not Use it or Continue to Deny It?

This is a question we must ask if we are to survive

"The group mentality and denial may be more than the psychological phenomenon of group denial where everyone is just nervous about stepping out of the crowd, but a more structured "Orchestrated Group Denial" in which anyone saying anything different is blacklisted, suffers career curtailment, threats and perhaps worse."

– Mary-Sue Haliburton,
Sept. 2008

"What we need is a truly free market in which people have the right to create and manufacture devices and products with zero or safe emissions, and the public has the right to choose freely to buy them and use them without hindrance."

– Mary-Sue Haliburton,
Nov. 2007

In a world gone mad with mendacious domination and greed, how can we as a species possibly get a grip on the problems that threaten our very survival? How can we truly solve global warming, global dimming and catastrophic climate change? Clearly, we must drastically curtail our incineration of hydrocarbons and research viable alternatives.

Some environmentalists are beginning to resign themselves toward weak solutions or nonsolutions such as drastically reducing our energy consumption while desperately increasing the use of solar, wind, hydrogen fuel cells, biofuels and nuclear power. The whole environ-mental movement is fractured into discord because no viable solution seems to be forthcoming even if we were to magically stop dumping carbon and particulates into the atmosphere.

To summarize our perilous situation: the routine burning of coal, oil, gas and uranium is creating climate instabilities many of us plan-etary scientists conclude *will* lead to global catastrophe if we don't do something about it soon. We have played God with our planet for too long. Yet our politicians, corporatists and media pundits keep fiddling while the Earth burns.

As a result of this gloomy outlook, some environmentalists such as James Lovelock and Stewart Brand have resigned themselves to nuclear power, in spite of the fact that it is unsafe (remember Chernobyl?), polluting (what do we do with millennia of radioactive waste?), and lead to the manufacture of doomsday technologies that are now spreading throughout the world. Nuclear "hot" fusion Tokomak reactors have proven to be unfeasible, radioactive, and too centralized to be practical. We have already spent tens of billions of dollars over decades of trying. It's a boondoggle.

We have also seen that other "solutions" such as sequestering car-bon or injecting particles into the atmosphere to offset greenhouse warming are band-aids that can only lead to further instabilities in our already-overburdened environment. I for one do not want to terror-form the Earth.

So we will need to go outside the box for our solutions, free of vest-ed interests and geared towards creating a sustainable future. With every passing moment, this vision slips away from us as we jump off the cliff lemming-like and the privileged ones amass an ever-greater share of power. These "leaders" prey upon our ignorance, confusion and inac-tion. They continue to divide and conquer the rest of us.

Economic arguments such as Peak Oil, while supporting our overall need to become unaddicted to hydrocarbon fuels, also reward the oil giants as they reap ever-greater profits because of the percep-tion, whether true or not, of increasing scarcity in the face of increas-ing demand. The Peak Oil issue seems to be a distraction, because macroeconomics itself is a contrived "science" that pales in impor-tance compared to the environmental destruction itself.

As I mentioned before, we are dealing with physical problems that demand physical solutions, and by process of elimination of con-ventional alternatives and ignoring the possibility of innovation, we seem to be facing our demise from destructive practices.

We need an energy solution revolution. We need an entirely new perspective. As in many other instances of profound social change based on necessity, most of us are afraid to think anew. We are looking backwards when we also need to move forward. Instead, we should be looking at the broader picture and seeking ways to allow innovation help us rather than destroy us.

The issue of what choices we must make on which source(s) of energy we want forces us to consider the issue of wise management of future energy policy, *regardless of our personal or institutional preferences.* It is something we will have to do anyway. Put differently, if we were not to get beyond our denials of new energy, it might some day be imposed on us in the form of weapons and centralized control by the cabal that is now in charge of our disastrous use of hydrocarbons and nuclear energy. Meanwhile most of the rest of us stand idly by, further abrogating our responsibilities as citizens.

In these times, the core of democracy within the U.S. has fallen to new lows. The awakening we need mandates an open discussion of the full range of energy options. Meanwhile we should oversee the orderly removal from power those who have led us astray, who have lied, suppressed and profited at the expense of all life on Earth. Fortunately, these forces are beginning fall under their own weight. But we still need to be prepared with the newer options to fill the coming void.

We need to awaken not only to new energy, but to reasserting our power, our intelligence. This is literally a case of "power to the people."

But we also must let go of our preconceived ideas and promotions, roll up our sleeves, do the research and development Apollo-style and have the democratic discussions. This is the only way I see we can move forward, even if your perception may be that new energy is, at best, a longshot. Or, after reasoned discussion and debate, if we willfully and respectfully reject the new energy option, we should know why, because such a decision might be to our collective peril.

So here is the multiple choice quiz that might just mean we re-inherit or dis-inherit the Earth: Should we: (1) use, (2) abuse, (3) not use or (4) continue to deny free energy?

The choice is ours to make. Let the debate begin! In the following sections we'll explore some implications of each choice to further frame the issue.

Using Free Energy. In my opinion, this is the only clear choice, after exhausting all other remedies, which is a process many of us in the scientific community have studied over the past several years.

The recent works of James Lovelock, Richard Heinberg, John Holdren, Nathan Lewis, Adam Trombly, myself and many others have all made it clear that there are no truly sustainable solutions using currently available technologies when full life-cycle environmental costs are considered. We seem to have no other choice but to embrace breakthrough energy.

This then forces the issues of responsible management of the new technologies, and setting the parameters of the transition to a new energy economy. To do otherwise would be to abrogate our responsibility as proactive world citizens.

Abusing Free Energy. One choice we can make is to again allow the U.S. Government or big business to control new energy. I think most all of you reading this can understand the grave dangers of this scenario. Our experience with the abuse of power by the existing war and energy industries and lobbies give ample evidence that free energy will be abused if we don't create whole new management and regulatory structures. I suggest that this issue will stand over all others once we begin to restore sanity to leadership and put it on a more global, peaceful and sustainable footing. There must be a ban on free energy weapons. Better still, we need a universal ban on all weapons of mass destruction: nuclear, chemical-biological, in space, and eventually, conventional weapons.

In other words, it is not enough to transform our tyranny in the ways the progressive community suggests. We must do more than that: to become intelligent about guarding the R&D for the common good, to guide us through the transition. The most difficult task of all will be to remove from power the vested interests while serving the larger interest. The issue of responsible application of free energy forces all other issues such as the what, where, when how this should unfold. This will take time and responsible citizen action during the inevitable chaos of making the necessary changes. Some vested interests will "lose." The whole capitalistic system and the concept of economic growth will have to be altered. At this writing, the system itself is imploding financially and morally.

In a sense, new energy is already being abused with the violent suppression of the technologies to date by the powers-that-be trying to squeeze optimal profits from the disastrous abuse of carbon-emitting and radioactive combustion that is throwing our planet so far out of balance. To keep these new developments secret is a crime as egregious as going to war for oil, having nuclear programs, ignoring the signs of climate change, and raiding the public treasury.

We must do more than restore the republic under the U.S. Constitution and disband the Empire as it stands. We need a global green republic made up of accountable leaders combined with regional and local autonomy, all aligned to the greater goal of providing clean energy for all peoples of the Earth without undue profits by the few. We not only need a housecleaning, we have to rebuild the house. The task is daunting but not impossible. By definition, these issues must be addressed by founding mothers and fathers of a whole new world.

These steps will be obviously difficult to achieve, and need to be central to our dialogues when we plan a new energy policy. Shaun Saunders wrote:

"Once the Pandora's Box of free energy is thrown open, how would this be achieved? And it is the threat of this misuse that could even persuade the eavesdroppers (i.e., secret government) that they are doing the right thing...my point is that the same human innate frailties that allow us to be so easily persuaded and entranced by consumerism (self-interest, the evolutionary 'need' to maintain our sense of status and importance, etc.) could also lead to our undoing in this scenario of unlimited free-energy for all: like letting the kids into the lolly shop, but with more profound consequences than sick tummies. At the risk of sounding like I'm an advocate for the very system that has caused the inertia referred to, as a consumer psychologist, I'm starting to think that freely available devices might be problematic. So, how can we introduce free energy—which IS needed—whilst ensuring that it is 'kiddy-safe'? It's almost like a classic twilight zone episode...How can we bring about individual and social maturity and regulate free energy (almost a complete contradiction in terms), and do so before the environment and social order have become damaged beyond the point of repair? I almost wish that there were 'powers behind the scenes' with an agenda that leaned towards solving this problem, as opposed to preserving the current energy-profit status quo."

The bottom line is that humans must manage the emerging technologies with a high degree of compassion and support for the entire human race, and not just the privileged few.

Not Using Free Energy. After due discussion and deliberation, there may develop factions who fear that inevitably free energy will be abused, and therefore shouldn't be introduced at all. Maybe we will decide that all technologies ranging from hydrocarbons to nuclear power to the new energy concepts are incompatible with nature and that the human experiment has already gone too far. Or perhaps we will decide that no new energy technology is truly benign and that we need to focus on improving efficiency and researching cheaper solar and wind technologies.

These issues are too important to be swept under the rug. Remember too that if you were to vote "no," free energy will almost certainly come to pass anyway. It's all the more important to closely track these developments than try to ignore them, especially given that the controllers aren't working in the public interest.

If the decision is a "no" to new energy, it will have to be made by a consensus of informed citizens and not by the current dominant powers. For that, the overhaul of global energy policies will need to be total, going to the roots of the disease. So inevitably the question keeps returning to the roots of human systems, not to the technologies themselves. We need to become educated and make responsible decisions.

Denying Free Energy. Even though history again and again confirms the Bertrand Russell maxim "The resistance to a new idea increases as the square of its importance," it still astounds me that the bulk of otherwise intelligent people still deny the reality and potential of new energy. That denial plays right into the hands of those who profit from Earth-destroying energy technologies.

The denial of new energy appears to be a defense mechanism to avoid reality because of a fear of the unknown. But more than ever, these times call for stepping up to the plate and considering all possibilities even though they may at first appear to be impossible. I know from years of direct experience, observation and study that denial is not only ignorance, it is covering up a truth that will inevitably emerge and be ever harder to face as we plunge ever further into global catastrophe.

Again, we can't sweep these issues under a rug of denial. The genie is out of the bottle.

Summary. When we look at conventional and new energy options, several political and economic factors arise. There is no question in my mind that we must squarely face the question of responsible management of the new technologies. To continue to deny it or to secretly suppress, develop and weaponize it are poor choices. But if we carefully assess these technologies and their implications, embracing the full range of alternatives, free of vested interests, we have a chance to transform our planet back to the paradise we had inherited.

Some of this chapter is based on an essay posted on July 19, 2006

Epilogue to Part II
Decisions, Decisions

"Don't be so open-minded that your brains spill out."
    – unsolicited advice I once received from a skeptic-colleague

"A stitch in time saves nine."
    – old proverb

"Decision, decision, decision!"
    – one bird's song where we live

After a long life so far, I believe I have finally learned the skill of nonviolence towards others. I can now withdraw rapidly from upwelling arguments, I erect boundaries between myself and angry people, and I stay away from places where the potential for violence is greatest (e.g., the USA, war zones and large cities). I like to stay at home or teach and retreat at yoga ashrams. Sometimes I speak at peaceful protests. Yet often I can't learn when or how to say "no" to well-intentioned inquiring people who come to us and unwittingly take advantage of our good nature.

So for all my restraint, I have sometimes forgotten to set boundaries. Every once in a while, I have also forgotten not to be violent to myself. Recently, many types of people have come into my email box, our village, and to us from North America to seek our counsel. Some of them are fed up with the hubris of the States and are personally struggling about what to do about it. They wallow in a desperate ambivalence. Some are intense, others are travelers just scouting or are on a casual vacation, but curious. They all look up to us for painting a scenario for what-could-be in their lives as new expatriates.

Most of them visited us in their own timing and for a short stretch of exploration. They came around with a thousand questions about settling here as an alternative to the States. Generally, they talked about themselves most of the time and had little curiosity about us as people, although they did appreciate the beauty we've helped create here. We expected nothing in return.

When about the fiftieth person came to us in this seemingly endless procession, and after we had held a couple of unexpectedly large and successful gatherings, I began to sense stress building up within me. I guess you could say I had never intended to be a full-time email correspondent, travel agent, tour guide, hotel-keeper, real-estate person, therapist, advisor and sociable guy almost on demand—especially when I'm writing a book. We ourselves were already busy building and settling into our new home and future retreat center and gallery; it's just that we came here earlier than the others seeking us out. We're simply more knowledgeable about living here, yet we were busy moving into our own visions. It became time to stop the influx and go back into my own center once again, regardless of what the outside world might ask of me.

So one evening, over a dinner we had prepared for one of the last individuals of the procession, who was a perfectly fine person checking out the area but arriving a few days later than expected, I began to split my personality (and then my head!) between being a gracious, witty and informative host and wanting to retire. I had a bit too much wine that evening (my Irish heritage plus a deep desire to explore different spiritual zones sometimes haunted me). Excusing myself, I got up from the table and took a few tentative steps away, when an arthritic knee buckled and I fell head-first straight into a cement pillar and then onto the hard tile floor in a pool of blood. K.O. to the count of ten!

When I came to, and with Meredith's help, I crawled to the bathtub to cleanse my wound, which turned out to be a long gash from forehead to the top of my head.

For the whole next day in my feeling of quiet embarrassment, I nursed my wound, which was mostly hidden beneath a mop of white hair. I began to realize I was lucky to be alive, let alone have no internal bleeding or mental dysfunction (that I know of: skeptics, here's your chance to discredit me once and for all!).

But I still refused to go to the ER (we avoid hospitals whenever we can). It was only on the following morning that I began to discover the wound was not healing well and I should go to the ER and get stitched up. This was a case in which hospitals were good, after all.

The doctor gave me a well-deserved lecture that I should have gone right away to the ER so we could minimize the chances of infection. He gave me nine painful stitches. Leaving the hospital, where there were no charges for the procedure, and looking like a gracious Frankenstein monster, I was hardly in stitches about this experience. I bought the usual expensive antibiotics and pain pills and returned home with lots of food for thought.

I asked myself, how do we make decisions? Why are they so often irrational and blatantly wrong? Enter my colleague and writer of the Foreword, Shaun Saunders—as if on cue.

When Saunders, himself a psychologist, was looking at this manuscript, he reminded me about the traps the human mind can fall into over individual and collective decisions. This, then, gave me an important missing piece about why we won't embrace the free energy option. Memories of my college psychology came flooding back to me about resolving conflicts.

Saunders wrote to me that when decisions between things we perceive are tangibly positive or good, the process allows us to thoroughly research the matter with pleasure. For example, most of us enjoy deliberating about whether we should buy the new car or the bigger television set. In our consumerist culture, we can simply drive down the street to the showroom, where eager, smiling salespeople are waiting to serve us. We cap our experience by showing off our new product to family and friends. It's positive reinforcement all the way. This is an example of what psychologists call an approach-approach conflict.

But when the choice is between negatives, Saunders wrote that we tend to freeze up and not want to find out the facts. Instead, we procrastinate and defer to authority. Should I go for chemotherapy or surgery? Let the doctor decide. Should we go to war for oil? Even if I may not like it, let the state and war industry decide. Should we destroy nature or go for the radical innovation? Let the big corporations that employ us decide. If we were to choose the new course of action, could we not risk losing our familiar livelihoods because of our vague fears of the unknown? We could cautiously proceed later maybe, but only if there is proof of demonstration and someone delivers it to us, or perhaps if I hear about it from CNN and if the president, Al Gore, GE or MIT approves, etc. etc. ad nauseum.

What about economic growth? What about bailing out predatory capitalists? We should let the authorities decide that one too, or at least we should postpone any radical changes in economic policies. After all, it's best to keep things stable; we don't want to rock the boat. Especially in these hard times, I don't want to lose my job.

These are examples of avoidance-avoidance conflicts.

Just like my (mostly unconscious) decision to be violent to myself and my ill-informed decision not to go immediately to the ER (this time, the doctor was right), these kinds of denials and procrastinations could have also affected our visitors who had contributed to my stress. They often grieved indecisively about the risks of leaving the old country, family and friends behind, and they also lamented the risks of the unknowns about coming here.

When some of them leaned into coming here, they often pulled back, meanwhile seeking our prolonged support while they wallowed in ambivalence. They would go back and forth physically or mentally, restlessly seeking a balance. *

---

*Some inhabit a netherworld of prolonged procrastination and just want to "hang out" in village cafes like Ernest Hemingway's "lost generation," waiting for something to happen to bless them: I can empathize; I've spent much of my own life like that. Some of them can gaze at us while we try to go about our work in their presence, which may appear to be social snubbing. Other people who have decided to move here, come back for more orientation. If they are needy or competitive, they often try to promote their own businesses and their own visions about community to us. They then expect us to respond and reciprocate positively. These folks on the surface can act confident, but express subtler and deeper manipulations of exchange with us that only come out much later when they project their disappointment onto us when we pull away.

It then occurred to me that many of these visitors were suffering from "double approach-avoidance conflicts," where the closer they were getting to their goals, the further away they felt pushed from achieving them—like buyer's remorse. Psychologists know this one well, and we became unwitting witnesses to our visitors' demises and projections. We shouldn't take this personally.

The common denominator of people's avoidance of decisions is that they are poised between two scary and unknown choices and are not taking responsibility for their own decision-making. By the way, many of our visitors are truly self-starters who appreciate our services as we do theirs—in very brief doses. They end up being our best friends and social equals.

Our collective decision about whether or not to pursue free energy involves the same kind of perception that the choice must be between destroying the biosphere and developing a disruptive and intangible outcome. Most all of us postpone the decision and addictively keep polluting. But when we lean into the positive possibility, we can again get scared away in a double approach-avoidance tug-of-war.

As I've often stated, the question of free energy is more a social-psychological one than a technical one. While the technologies themselves wait in the wings, we all go through an irrational and emotional process of decision-making or indecision-making in our own minds. Few of us want to take personal accountability for being a part of that decision. Saunders and I both believe that if we confront these decisions as responsible individuals in search for the greater good, we have a good chance to break through. But many more of us need to do so. This can be a lonely process.

For most of us, addressing the free energy decision is a very long journey, perhaps too long to get the results we want. Wade Frazier identified at least eight major steps we must endure to "peel the onion of free energy." (www.ahealedplanet.net/paradigm.htm)

I believe the most important step we can each take is *education*, finding the facts of the matter regardless of prior beliefs. But the process always seems to require our going through unpleasant emotions that put us "between a rock (biocide) and a hard place (fear of the unknown)." There are few short-cuts we can take through the emotional phases of grief and transformation, as I learned in my research for *Miracle in the Void*.

But to educate ourselves and one another, we each need to take personal responsibility, and then join up with our own kind. I thank my friends Shaun and Wade for pointing this out to me at the most important moments. I also thank my higher self for knocking some sense into me.

In the next part of the book, we will shift the focus towards asking each of us to go deeper into how we can intelligently address the most ignored and most promising options outlined in the last few chapters. What do we need to do about the decisions that lie ahead? Let's be honest: what kinds of energy sources *do* we really want? Should we use, not use, abuse or keep denying free energy? Most importantly, how can we *as individuals* stand up and be counted for taking a road less traveled? What choices would *you* make? How can we effectively organize a new world of sustainability and abundance?

Act as if it's your choice, *you* decide. We can't expect, nor do we want, others to do that job for us. If you must, imagine you are president (sic) of the world. But no matter what, please don't expect or even want others to do that job for you.

We have some birds near the house who chant with impeccable enunciation, "Decision, decision, decision!" The choice, dear reader, is yours.

Part III

We Need a Transformation!

Chapter 18
Unifying the Four Cultures of the
Phoenix: An Experiment in Democracy

"The entire planet is now unstable because politically "correct" people have continually watered down the message and therefore the impact of the message regarding the Emergency on Earth. Some people, it seems, think that the message regarding our collective and extremely dire situation can be diluted to accommodate the sleepy. Unfortunately for us all, the Earth is not polite in her response to our abuse…we stand on the deck of the *Titanic* which has already hit the iceberg and is definitely sinking. The majority of our fellow passengers are trying to protect their deck chairs, while drinking their anesthetizing cocktails and gossiping as if nothing is happening at all…I have spent the vast majority of my adult life warning the People of the Earth about what they are creating while the majority look the other way. I have always offered real solutions and not fake ones only meant not to offend those condemned to ignoble deaths at the hands of the international corporate hegemony."

– Adam Trombly, www.projectearth.org
June 14, 2008

"Economic growth this century could be 32 times as big an environmental issue as population growth. And if governments, banks and businesses have their way, it never stops. By 2115, the cumulative total rises to 3,200%, by 2138 to 6,400%. As resources are finite, this is of course impossible, but it is not hard to see that rising economic activity—not human numbers—is the immediate and overwhelming threat."

– George Monbiot, *The Guardian/UK,*
January 28, 2008

O ver fifty years ago I began to explore the idea of the separation between the arts and sciences, since I loved them both and I wanted to do plenty of both in my life. In his famous essay written in those times, *The Two Cultures*, C.P. Snow lamented the cultural disconnect between scientists and artists. That work helped inspire me to follow a liberal arts curriculum and to blend the two. But as life went on, I could tell I was fighting a losing battle. Career pressures put me squarely into the institutional scientist camp until about 25 years ago, when I finally declared "enough" and I left, pursuing a more integrated life. The division of cultures embedded within the same society is a question that has always fascinated me.

Thanks to our recent animated dialogues, I saw cultural divides happening in new ways during the June 2008 Phoenix gathering, a dynamically-facilitated democratic experiment among 25 motivated participants all passionately committed to finding "sustainable solutions to the crisis of civilization." (www.wakingthephoenix.org) The participants, many of whom are known for innovative thinking, traveled from all over the world to our peaceful new mountainside Andean retreat, Montesueños. They came from Ireland, India, Argentina, Canada, Mexico, the U.S. and other parts of Ecuador, for a week of intensive discourse. Our far-ranging dialogues began to pose for me some deeper questions about the relationship between a given culture and sustainable breakthrough possibilities. Why do many of us deny and resist them so? What cultures do we really represent? Why do we have such difficulty converging on solving the overarching problems that we all feel so passionately about and that brought us together? Why do we too, as a democratic group, tend to water down the solutions we really need to embrace?

I began to find answers when I felt taken aback when one professionally successful participant calling himself a pragmatist shared his mental stress from what he saw as my inadequate explanations about free energy research. This encounter for me once again raised an important set of issues about the epistemology of breakthroughs in our democratic discourse coming from a clash of cultures that aren't really talking with one another.

In my case, I gave a well-publicized-in-advance lecture one evening during the gathering, which addressed the state-of-the-art of the free-energy research. Some of the group came to the talk, but our pragmatist and some others didn't. Even so, this is not a simple subject to comprehend at first blush—not only technically but culturally. It has taken me twenty years of deep study, including world travels to the innovators, skeptical analyses of the concepts, and a lot of personal research and writing, to begin to embrace the enormity of the potential of breakthrough clean energy. So for me to expect him to "get it" or for him to expect to receive this information from me concisely and convincingly was unrealistic, and of course, this was not the first time my presentation was not well-received.

This evident failure of communication I'd experienced so many times presented a unique opportunity for me to learn about where we all come from and how to communicate better in the future if we can stay with our commitment to dialogue—a powerful part of our experiment. Most pragmatists would have run for the exits and kept walking. Unfortunately, after the cloistered intensity of the gathering itself and everybody went back home, the pragmatists did walk and never came back.

Given these practical challenges to being able to learn about the possibility of free energy, I would prefer that skeptical pragmatists suspend disbelief and address the general question of how technological quantum leaps might benefit our future world and also raise questions about how we can regulate them. There is plenty of reference material on free energy in the peer-reviewed scientific literature and on the websites www.brianoleary.com, www.ahealedplanet.net, www.newenergycongress.org, www.infitite-energy.com, and many others.

Our group pragmatist prototype shared his opinion that some belief-systems of a few of the others in the group, sometimes condescendingly, because they did not match his own beliefs. I prefer to call these systems "cultures." Within our group, I identified four such cultures as a convenient point of departure for understanding our differing approaches to our central concern. The gathering included each of these radically different cultures—only one, possibly two, even three or (rarely) all four of which each of us can feel comfortable with. In respectful consideration for each culture, I describe them as follows:

1. The so-called Paranoids (aka, conspiracy theorists) might be better described as *Truth-Seekers* who address unconventional breakthroughs and the lies, cover-ups and deceptions by vested powers that are blocking the new possibilities. These suppression and disinformation campaigns are well-known to those of us who have been victims, but the information is also available in the alternative scholarly literature, clearly showing many promising

solutions such as free energy research that have been stopped. The best antidote to the secrecy is coming forward with realities in ever increasing numbers as truth-tellers. That this culture has been frowned upon represents a deep cultural divide. Embracing unpopular truths can be a risk to one's career—and even to life and limb. For years, I sported a bumper sticker, "The truth will set you free but first it will piss you off."

2. The Doomsdayers might be better described as very concerned *Deep Ecologists* knowledgeable about the extent of the destruction upon Earth wrought by Man. I believe most of them do not exaggerate. The prognosis is grim unless we radically change our ways. But these people are often afraid of new technologies, because of humankind's horrible track record of their abuse. This culture is also frowned upon, because the news the Deep Ecologists present is too sad to bear. Most of us are in denial about the bad news, or we'd rather shoot the messenger. Addressing these overwhelming issues can especially affect the most sensitive among us, and has led to the emerging field of ecopsychology, which is designed to move us through our grief about the death of nature, so we can become empowered to do something about it (e.g., see *Re-Inheriting the Earth*).

3. The *Spiritualists, Consciousness Researchers and New Scientists* are generalists who are also profoundly concerned and feel that appealing to a higher consciousness will help us unify in our mutual quest. Scientific experiments show, for example, that combined positive human intention can create significant results in the material world, e.g., healing, prayer and purifying water. Some scientists have also found evidence for life-after-death, contact with higher beings and the profound interconnectedness of all life guided by the spirit of Gaia, or Mother Earth. Many frown upon this group because of the Spiritualists' beliefs in "magic," a transcendent reality that seems to scare off the secular pragmatists, whose main beliefs are rooted in materialism and reductionism. From my own and others' experiences with yoga, meditation, being in nature, and other spiritual practices such as indigenous ceremony, many of us have learned that these approaches can be very powerful—both individually and collectively.

4. The *Progressive Pragmatists* (PPs) are people such as our own

group example. These individuals want to find a consensus and solutions within well-defined parameters consistent with the kind of professionalism that goes along with a career. The PPs are reform-oriented within the current frame of mainstream culture. I consider myself as having been a pragmatist as an Ivy League professor until about 25 years ago. Only then did I become educated about the extraordinary degree of our challenges and the possibilities of out-side-the-box solutions. I could tell from years of dialoguing with the PPs that it is a long process for a PP to begin to appreciate the impor-tance of the contributions from the other three groups. In fact, as soon as I made some shifts in my own thinking, the cultural divide between myself and most all Pragmatists became enormous, much more so than I could have imagined. I also remembered having been on their side of the divide. I too had scoffed at the Truth-Seekers, Deep Ecologists and Spiritualists, because it was a politi-cally correct way to behave with my scientific colleagues.

By later embracing the other three cultures, I suddenly felt liber-ated from the straightjacket of "pragmatism" but I also felt rejected by those still ensconced in the old scientistic culture. To them, I could no longer practice science—*their* scientism, as if my scientific skills were suddenly null and void. I felt I suddenly "disqualified" myself as a "credible" source to the Pragmatists—for example, when they ignored me or placed the burden of a facile proof-of-concept on me rather than if they just mustered the discipline to do some learning for themselves. It was only then I discovered that this pattern of denial of the new has happened with paradigm-busting breakthroughs throughout history. That's a pity, but the very fact that we were dialoging about this was an encouraging sign.

I sometimes call the Progressive Pragmatists "the pushers," tak-ing the current Western paradigm and expanding it as the most realis-tic approach to the huge problems presented. In today's world, the Pragmatists, liberal and conservative, are clearly in charge of our mainstream secular institutions. For the most part, they focus on what is in front of them rather than spend time on long-term visionary thinking. Usually career trumps shifting the paradigm. The tempta-tions of power and money can distract the Pragmatists in particular. These dynamics do not apply as much to the other three cultures. Outside of cults and organized religions, the Truth-Seekers, Deep Ecologists and Spiritualists are generally freer to express and there-fore constitute "the pullers" toward a larger worldview.

In my opinion, we need all four cultures for a diverse group such as ours to achieve significant results. In that respect, I think our gathering was a great success if we were to further open the dialogue. In principle, more closure could follow once we appreciate more fully where each of us is coming from. We can indeed be from very different cultures, yet still be able to address the very same mega-problems we all feel so strongly about solving. But I'm afraid that was wishful thinking, because the split among the cultures seemed to be more important than our collective willingness to look at radical solutions. I learned by experience that well-intended democratic discourse even among progressive thinkers was not enough to give us the jump-start we needed. It seems that the scattered and unorganized thousands of us willing to embrace free energy issues need to have great tolerance and understanding of all four cultures.

On the other hand, some progressive pragmatists can also create (incremental) miracles of potential change. Representatives of the Phoenix group recently met with the Ecuadorian minister of finance, and some brilliant proposals were presented: the practical, economical preservation of the precious resources of the oil-rich and biodiverse Yasuni rainforest through the sustainable use of herbal and biotechnical medicines possibly worth more than the oil there; the use of alternative currencies for Ecuador to brace against the collapse of the dollar; and the application of re-localization ideas. This kind of interchange is essential to bridge the gap from the ignorance of nowadays towards a bright future for a nation in re-formation—and for the world. I applaud and support the efforts of Bruce Cahan, Chris Shaw and other "pragmatists" in our group for providing these bridges toward the future.

It's my belief that breakthroughs will determine our future more than a consensus of any committee that won't look at the coming quantum leaps and are culturally separated. Truth-telling, free energy, planetary healings, and possible catastrophic biospheric, political, economic and societal collapse and are real concepts that were acknowledged by the plenary group but not really seriously discussed. It seems hard to get beyond the questions: Are these breakthroughs real or possible? Should we be serious about considering these as viable disruptive shocks and solutions if most mainstream scientists, media and "environmentalists" don't even give them the slightest attention? Therein lays the cultural divide between the Progressive Pragmatists and the other three subsets. We need to acknowledge this cultural conundrum in future dialogues and seek ways to bridge the gap. We need a greater blend of cultures that could lead to a higher degree of sentience to embrace free energy.

The largest common denominator of the Phoenix Gathering was the re-localization movement. We all recognized the tyranny of powerful nations and corporations should not rule the world; instead, local governance and enterprise were far more desirable. One of the five subgroups in which I was involved was called the Innovation/Free Energy group. We came to the conclusion that innovation could be fostered in safe, sustainable regions of Earth that would support the efforts, free of political and environmental pressures. These protected "incubation nests" would hopefully prosper, even through difficult times, with the potential to provide the world with free energy, for example, or to at least keep the torch of civilization burning like the monasteries of the Middle Ages (e.g., see Morris Berman's prescient book *Twilight of the American Culture*).

Nevertheless, the Gathering participants discussed more about the where things should happen (re-localization) than what must happen on a planetary level (e.g., breakthrough energy, enduring peace, sustainability and international justice). We fell short of embracing the potential of the concepts themselves of free energy, consciousness, spirituality, combined positive human intention, water purification, and other innovative concepts, nor did we discuss forms of global green governance that could help guide those ideas that could apply to all of civilization. The thrust was instead local rule, going back to the farm, so to speak. But can we hide from the tyranny? Should we be secretive or transparent about our intentions to do this work?

Discussion of outside-the-box concepts leads to a deeper understanding of what democratic discourse about transformational ideas can lead to. It could unify our diverse cultural perspectives. This gets to the core of what we really need to do together, without re-inventing old wheels destined for another dustbin. We need to be bold. It would seem that our next step might be to trust more the knowledge and wisdom of those of us who have toiled for decades on some of these larger questions, to listen more to those of us who have a long track record and motivation to share the rich expertise that can come from each culture, and to get beyond the frustration some of us may have felt at times about the duality between the fact that while we are equals as persons, some of us know more about some things than others. That would be the next step, I believe, and the foundation of trust is an important first step.

Our goal, then, should be to seek unity arising from the very diversity of our cultures and to have the courage for us all to address the question of breakthrough solutions. If the Phoenix Gathering itself didn't achieve that goal, we'll need to gather those kindred spirits who have already embraced the possibility of transcendent solutions such as free energy. For that, we'll need to have a great deal of awareness of and compassion for our cultural differences and the importance of all four or more cultures in our mix. This mutual respect comes from the value each of us brings to the table.

At the moment, it seems that only some Truth-Seekers and Spiritualists are ready to try this one on, the Deep Ecologists can provide the incentive to do something about it, and the Pragmatists can provide the means.

In summary, the Truth-Seekers contribute by uncovering politically forbidden realities. The Spiritualists evoke a higher power and combined positive intention to see the big picture to help heal us and transform us. The Deep Ecologists give us the needed wake-up calls. And the Progressive Pragmatists, if they can only yield their own prejudices about reality, power and control, could join us in coming up with the necessary transition strategies for a new world. The search for the needed blend of attributes is on. We need to find the founding mothers and fathers of a Plan C to achieve a truly abundant future for humankind. Are you up to helping the cause of such an awakening?

This chapter is based on an essay posted in July 2008.

Chapter 19
A Call to Arms
(not the exploding kind):
A Personal View

"Every new generation needs a revolution."
– Thomas Jefferson

"The most dangerous person to any government is the one who is able to think things out…without regard to the prevailing superstitions and taboos. Almost inevitably he comes to the conclusion that the government he lives under is dishonest, insane, intolerable."

– H.L. Mencken

"The Earth is standing at the edge of a dark precipice; on the other side is the greatest epic in mankind's history, waiting to be written, waiting for those of us who dare to take up the pen and to fight the darkness of the lies. An army of freedom-writers, who are as dedicated to defending life as the enemies of life are dedicated to ending it, can bring forth a new sustaining vision of life. We must first choose to place our own freedom in jeopardy by stepping forward to stop those who would take us into the void in a vain gamble to crush most of the Earth's life for the sake of greater 'profit.'"

– Peter Chamberlain, www.globalresearch.ca,
Feb. 2, 2008

"Through my signature below I hereby withdraw my consent to be ruled by the organization that has called itself the Government of the United States of America. A government is empowered through the consent of the governed to serve a sacred purpose, namely to create a bright and sustainable future for its people and a biodiverse garden of its region. This purpose is possible. If a government no longer serves

its intended purpose then it is proper that each individual formally withdraw his or her consent to be ruled by that government."
                                            – Anonymous, www.globalresearch.ca,
                                                                July 3, 2008

"The prophets remind us of the moral state of a people: Few are guilty, but all are responsible...We must ask ourselves why we don't do more, while recognizing that none of us can ever do enough...We must be harsh on ourselves and each other, while retaining a loving connection to self and others, for without that love there is no hope...If we wish to find our prophetic voice, we must have the courage to speak about the crimes of our leaders and also look at ourselves honestly in the mirror. That requires not just courage but humility. It is in that balance of a righteous anger and rigorous self-reflection that we find not just the strength to go on fighting but also the reason to go on living."
                                            – Robert Jensen, Professor of Journalism,
                                                    Univ. of Texas, Austin, July 8, 2008

C learly, to achieve a future of joy and abundance for humanity and nature, will require a global transformation in our collective thought and intention, which must take place to nurture the visions of sustainability, peace, and new energy. Lester Brown's Plan B and Harvey Wasserman's *Solartopia* fall far short of the mark. We need to introduce a Plan C for discussions among those of us willing to look at it.

I propose this option with some remorse towards those with moderate or incremental views, and with the following cautions: I am of Irish descent and have fire in my belly about the atrocities and suppressions committed in my name, as a citizen of the United States of America, I deplore the crimes of both commission (illegal wars, deficit spending and election fraud, for example) and omission (neglect of environment, health, fiscal regulation and breakthrough energy, for example). The foreclosure of Plan C by even those advocating Plan B is part of the problem.

At risk of offending moderates, I depart from my normal sense of diplomacy to declare "The emperor wears no clothes!" We cannot get from here to our destination with a moderate agenda. It will have to be radical, meaning to go to the roots. I declare that the repressive oligarchy of my own country of the past decades, culminating with the policies of the Bush administration, can no longer be trusted to control our collective futures. I declare this from fifty years of experience with academic, industrial and governmental systems. I do not take my outburst lightly. There needs to be a public outcry if we have any hope for peaceful change. I also invite feedback from moderates, who may or may not agree with me.

Since the beginning of time, humans have faced the ravages of war, political corruption and ignorance. But never before in history are we so poised on the brink of planetary ecocide. Those in charge have gotten away with the most massive show of greed and aggression the world has ever known. These misguided "leaders" have no interest in nurturing the true visions of planetary sustainability. They have seized our collective sovereignty and are creating an illusory war footing pre-empting any kind of global transformation needed for our own survival.

Sadly, the rest of us have allowed that to happen. Lemming-like, we are poised on the cliff-top seemingly wanting to fall and perish rather than to face the truth of the deceptions of our enemies, both from within (complacence) and without (warmongers and resource exploiters for profit).

One significant force is the disproportionate influence of those with wealth and power. This dynamic again suggests we may need to have another "revolution" such as those experienced in America, France and Russia in past times. I'm sure most of us pray for a peaceful coming transition. But what we really need to have is a transformation from our current scarcity-based paradigm to a benevolent, clean and abundant one. We must emerge from our collective sleep, at first in small numbers of the most aware among us, and growing.

I am deeply sorry for what my native country, the U.S., has done to the world and wish I could have done more about it earlier. I had marched on Washington reporting to the White House staff my outrage about their secret invasion of Cambodia in 1970 (next chapter). But I still hadn't learned my lesson of political transformation: that we must sustain our protest, articulate alternative views and massively mobilize—not just come out one day to demonstrate against a war. In the near term, the new agenda of Plan C will be framed by small numbers of us who are searching for one another as kindred co-creators.

I reluctantly conclude that the old ways won't work. We must now step outside the box of our collective mendacity and create a peaceful transformation, to pull out all the stops, to open up to solutions. We must shift our focus away from the destructive follies of the leaders of the American Empire and their corporate cronies. We cannot any longer tolerate their conquest for oil, their weaponization of space, their slaughter of innocents, their privatization of nonrenewable resources, their rape of a global environment in its eleventh hour of stability and nurture, their obscene defense budgets, their nuclear saber-rattling, their outrageous lies, their suppression and secrecy, their robbery of the public treasury, their briberies, their tortures and assassinations, their fixed elections, their accelerated massing of scarce resources, and their total disregard for the dignity of humans and nature and our collective future. That is not a government which in the interests of the people. It is a diversion to fascism. We need to recognize this for what it is for us to make the necessary changes.

And perhaps their greatest crime is our leaders' denial of international cooperation necessary to reverse global climate change, crony capitalism and the gathering war machine. I wish the leaders could just apologize and step aside and help us make the necessary changes. My 2003 war protest placard had said, "Exile Saddam, exile Bush et al." My more satirical 2009 placard might read, "Nuke 'em, go for the oil (snort) and pray for the Rapture." My real-life 2009 bumper sticker reads, "Keep the oil in the soil: develop free energy!"

We cannot any longer allow the excesses of polarized forces to lead us into the abyss.

"Give me liberty or give me death" were Patrick Henry's stirring words during the American Revolution. I repeat those words in a somewhat new manner: "Give me liberty, give me global sustainability, give me peace, or give me death." For my own coming physical death (which I do hope will be natural) will be nothing compared to the premature death of all life if we cross the perilous threshold awaiting us and continue to allow criminal ecocide and total war to dominate our lives—just to serve the interests of a controlling elite.

We must awaken to new ideas on a global scale. We should look the Empire in the eye and collectively say, "No, we won't play your game any more. We do not recognize you as our legitimate government." The fall of the American (and British) Empires must lead to an entirely new form of global governance. We do not have the luxury of time on this one. We must now step outside the box and let new ideas guide us. We can either create a renaissance or a disaster. The oil age is ending, let's get on with innovating. Yet my words still seem to fall flat when it comes to this truth, and so, in these times, we can only reach out to kindred spirits who "get it" that both our problems and solutions need to be addressed from a very deep place.

What remains to be determined is how we can restore humanitarian principles for our collective survival, and do it with love and compassion. Free energy lies at the very core of the needed changes. We are encouraged enough with the wide range of cold fusion and vacuum energy results to band together as a people to support the needed research and development to make new energy a reality. It is clear these developments are perceived as a threat to the prevailing powers. We must find ingenious ways to develop and distribute this technology before it's too late. We must offer the people choices of energy sources that are feasible, cheap, clean, safe, decentralized and publicly transparent. I am convinced we can end this nightmare of war and pollution.

The basis of my convictions is years of research and visits to the innovators of new energy and other breakthrough concepts leading to global sustainability. These ideas have been hidden from public view for more than a century, since Tesla's times. I have concluded that all we need to do is support the research and development of these technologies under responsible stewardship. I have also concluded that we have no choice but to do that, given our grim planetary prognosis.

We must start afresh. American politics has not only escaped reality, it has become the greatest obstacle to global sustainability, freedom and innovation. Iraqis defending their homeland don't hold a candle to the official violence of American and profit-centered war criminals, Republican or Democratic. Yet at the same time I do not want to offend the progressives or any other concerned citizens of our planet. I'm only saying that we will not be able to achieve our goals without a basically radical change in our perspectives, and the traditional political and corporate arenas are not where it can, should or will happen.

The real goals have slipped away from us: abundance, clean energy, environmental and economic sustainability, peace on Earth and in space, human rights. The answers might be given only lip service but no sincere initiatives have yet begun. Free energy leads the pack, fomenting the greatest technological revolution of all time. Yet we all know that the prevailing governmental or corporate powers have never wanted this. There's little money or central power in fostering this transformation. Yet it must happen for our own survival. And so we need to form coalitions outside the system, first in small numbers of courageous activists, to take worldwide steps to oversee the needed R&D and to hold responsible those who have betrayed the public trust.

New water, forestry and agricultural concepts also wait in the wings for their opportunities to create a peaceful and sustainable future for us all. But we must insist on this...or we all die. How many more martyrs do we need to see go to the gallows to awaken us to this fact? Zero (I hope)? One? (if so, I'd be happy to be that person). 100? 4000 U.S. soldiers? 200,000 Asians unprepared for the devastating 2004 tsunami? 1 million Iraqi civilians? 500 million orphans worldwide? 1 billion without adequate water? 6 billion? All life on the planet?

Yet since the time of Tesla one century ago, we are both closer yet ironically further away from our goal of new energy. Closer because the technology is now well along; the Wright brothers have already flown on this one many times over, in terms of proof-of-concept. But we are farther because these ideas have been systematically suppressed by the powers-that-be. Most of us are in denial about our overconsumption of nonrenewable resources. We seem to prefer plunging over the lemming-cliff without so much as a thought about the rocks below. And all we need to do is to do our R&D freely and transparently, a job that should have been done by the U.S. Department of Energy, which seems to be more interested in nukes and "clean coal." We must transcend that repressive consciousness which has so steadfastly and for so long held sway over us, while we vainly try this or that rearrangement of the *Titanic* deck chairs.

The old ways won't work. Appealing to genteel, progressive intellectuals, whose mental musings are valid but getting us nowhere. Appealing to the much-duped and brainwashed American people, won't work either. Yet the common people of the world can unite and arise peacefully against this great tyranny and offer solutions. These movements are beginning to gain political power here in Latin America.

We must re-invigorate the global community with new ideas. A leading conservative Allan Bloom said, "The greatest tyranny is not the one that uses force to assure uniformity but the one that removes the awareness of other possibilities." That is true. Our obstacles are not technical, they are based on human greed which suppresses new energy and "other possibilities." In the U.S. government's insane "War on Terror," it will need to look in the mirror. At some level we know what to do. We need a transformation built on coalitions of those who not only believe in progressive causes but who believe in the sanctity of life and that wisely-chosen innovations can help us create a new global agenda, a new prosperity for us all.

At the time of this writing, the constituent assembly of Ecuadorian citizens completed drafting its new constitution, which the people ratified with a strong majority vote. Included is an unprecedented provision to declare that nature has rights. This achievement could lead to a whole new set of principles ignored by all governments throughout the world, and sets a precedent to leave nature alone, to keep the resources intact, and to stand up to exploitive empires and oligarchies.

Meanwhile, we must sidestep the American government such as it is, and vigorously oppose its neo-liberal and neo-conservative policies. Bearing in mind the Bush administration was not legitimately elected on both occasions, it is easy for us not to recognize the U.S. government as legitimate either. The U.S. has been hijacked by unlawful bankers, crooks, warriors and polluters. What the media and others call the "U.S." then is not the U.S. at all. Our "president" is not president, but merely an actor playing a role in the service of fear and greed. We must sidestep corporate power lubricating this criminally suppressive machine and open ourselves to the new solutions, undistracted by the seething deceit and lies of our leaders and their media and academic mouthpieces. It's time to take back our power, forgive and move on. Let's coalesce with decent humans worldwide, the workers, the grass roots, to insist on a stable, sustainable future for the planet. We should network our agenda all over the world with those who have a sympathetic ear until we gain the support of most sensible and compassionate people.

Let us finally cast off the heavy yoke of tyranny under which we have endured for so long. Let us invite courageous new leaders and groups who could rise up in an unprecedented and massive transformation for new energy and for a real peace on Earth and in space.

I do not take these remarks and intentions casually. In the next chapter, we'll explore the history of contemporary American tyranny, as I personally and intimately experienced it in Washington, D.C. I've played many different professional and activist roles during the past fifty years.

Portions of this chapter are based on an essay first posted in June 2005

Chapter 20
Return to Washington D.C., September 11, 2006 and 2008

"To preserve their (the people's) independence, we must not let our rulers load us with perpetual debt. We must make our selection between economy and liberty, or profusion and servitude."

– Thomas Jefferson

(King George) is, at this Time, transporting large Armies of foreign Mercenaries to complete the Works of Death, Desolation and Tyranny, already begun with circumstances of Cruelty and Perfidy, scarcely paralleled in the most barbarous Ages, and totally unworthy of the Head of a civilized Nation…A Prince, whose character is thus marked by every act which may define a Tyrant, is unfit to be the Ruler of a free People…it is their right, it is their duty, to throw off such Government.

– U.S. Declaration of Independence

I. Fifty years of encounters with a city and nation in crisis

The nation's capital and I have had a five-decade-long, deep and emotional history. I suspect that many of you readers may also have stories to tell about the mysteries of your own relationship with our beloved and betrayed contemporary Rome.

From the sprawling lawns and marble buildings to the 100-to-200-year-old row houses in the poorest (black) and richest (white) sections of town, I have lived in Washington many times in many positions in many neighborhoods from suburban Maryland to DuPont

Circle, from Georgetown to Capitol Hill, from the slums of Northeast to the three-martini lunches with lobbyists and the bag lunches with environmentalists, all the while experiencing the mood swings of the seasons from terribly hot summers to icy winters to colorful, lush, sensuous springs and autumns. Each time I lived there felt like a lifetime.

For most of my life as a young adult and an adult, I have intensely engaged myself with Washington, D.C. What was most poignant about these experiences was that, more than anywhere else on Earth, it attracted career-types and power-seekers and a sprinkling of altruistic souls, or sometimes a curious blend of two or all three of them. Always. It's been said that if you're at a cocktail party in Boston, an introduction always concerns what family you're from and what prestigious college or university you attended. In New York it's about your investments, publishing, promoting, advertising and entrepreneurship. In Washington, it's about your career and position.

I know about this, because I was a career- and occasional power-seeker myself, constantly looking for a way to make a difference in creating a visionary, just, peaceful and sustainable future for humanity on the Earth and in space. But at the same time, my ideals had to be tempered by a kind of bland realpolitik that pervades the mythology of inside-the-Beltway thinking. After all, I had a career to protect and a living to make. During most of my Washington years, I had children to support and a mortgage to pay. It turned out it was this conflict between idealism and realism that led me to such angst, to a love-hate relationship I have felt nowhere else. I suspect this dynamic applies to some others too, in contrast to the kind of experience a well-healed lobbyist or politician may not be able to understand as an insulated courtier to the kingdom.

Washington for me was always a catalyst and reflector of my own aspirations as the nation's history and my own history weaved, bobbed and meshed into and out of a volatile stew. It happened first as an optimistic teenager, then as a college graduate and space scientist, as a graduate student and high school math teacher, as a NASA astronaut, as a college professor and Vietnam war protester, as a Star Wars and space shuttle boondoggle analyst and Senate testifier, as a Congressional staff consultant and speechwriter on energy end environmental issues, as an advisor to presidential candidates on space, science and clean energy, as an industrial contractor, as a space visionary, critic and author, and as a nonprofit advocate of the peaceful uses of space.

Each role I played, each position I held in Washington seemed to start with a hope, only to evolve into an entrapment and finally an exit which would only lead to new roles and positions months to years later—back in Washington! As I said, this relationship has always been deep and karmic and may even have been a codependent addiction.

Then, around 1990, I just as suddenly and unexpectedly removed myself from the scene towards what seemed to be a permanent divorce. My "last parting" happened when I stepped into unprotected wet concrete in front of a Senate office building during an innocuous brief visit during the early 1990s. As I was sinking into an oozing slow-motion descent of more than a foot, I briefly envisioned no escape. Like Washington itself, I felt concretized by vested interests that carry the day. The symbolism was poignant: at some level, this kind of entrapment would endanger my future, our futures and those of our children. I had no choice but to get out of there fast.

Moments after my clumsy and embarrassing self-extraction, a part of me also wanted to have left an imprint and I was tempted to stay with the action. But I didn't. It was a nice neat hole with my shoe well-molded but too deep to be safe for future pedestrians. But I could have created my own safe one and guarded it to form a permanent imprint.

The Founding Fathers had left their imprints by falling into their own high-risk ooze, extricating themselves from it, and then creating an imprint that has endured for over two hundred years. In my case I didn't stay long enough to create or protect my contribution: the efficient workmen came back to smooth over my deep and unsafe footprint and the sidewalk returned to business as usual. I wondered how often we might risk, consciously or unconsciously, stepping into wet concrete in order to make an imprint just to have it once again obliterated by conventionality.

The fear of the unknown is scaring us away from the imprints we so passionately need to make, but somehow we don't have the courage to take measures to fall into the concrete, then get out, and finally make and preserve an imprint of our own design. All three steps are needed to restore the republic and to redesign those parts of it to fit into today's world. In my hilarious experience with concrete, I only did the first two steps. That wasn't enough to make a difference. You should have seen all the serious people walking around me, some condescendingly gazing at my slapstick maneuver.

But I am getting ahead of my story. Later in this chapter, I'll explain why and how I blew out of Washington for more than 15 years as an activist. Call it outrage fatigue. But I was back now in 2006, just as suddenly, seeking to penetrate this unsettled marriage with my native land and its capital. I expect many of you can relate to the ambivalence and mystery of this relationship, because Washington symbolizes this thrashing giant of a nation made up of our collective power and greed, a nexus of our hopes and our fears. Right now, fear has taken over the national agenda.

The point I'm leading to is, my own history with Washington and every thinking person's relationship with the United States is filled with such possibility, disillusionment, complexity, contradictions, betrayals, narcissisms, obfuscations, power lusts, and for me, an emotional roller-coaster ride like I have experienced nowhere else.

For the sake of simplicity I divide my own history with Washington D.C. into the following six epochs: (1) my (our) optimistic patriotic high (school) of the post-World War II years, mostly during the 1950s, (2) the Camelot years of JFK and the Apollo program during the sixties, (3) the war protesters, the critical interventions, political aspirations, and visions of a peaceful, sustainable and futuristic expansion into space during the 1970s, (4) confrontations with neoconservative visions such as Star Wars during the 1980s, (5) outrage fatigue during the 1990s until now, (6) a return to "take back Washington" with kindred spirits who want to restore and improve the republic starting on September 11, 2006 until the present, and (7) a September 11, 2008 retrospective on the possibilities of a second American Revolution or whether the seeds for global transformation could come from elsewhere.

This chapter explores how my history and our collective history can mesh to create new visions of a nation now under siege, not by Muslim terrorists, but by our own criminal "leaders" housed in Washington, D.C. and their obsequious followers. They should not only be removed, we must also formulate an alternative vision that will facilitate the transition and to ensure our survival. To do any less would be to abrogate our responsibilities as citizens of the most powerful and sick nation ever on Earth.

First the history, then some insights based on my experiences and other research:

1. *The 1950s.* These were the years of the "The High" expressed by William Strauss' and Neil Howe's brilliant and prophetic book *The Fourth Turning*. We all liked Ike and felt a great promise and power of a victorious nation in peace and prosperity at last. This meshed well with the optimism and impressionability of a young teenager

climbing the stairs of the Washington Monument and gazing in awe at the Lincoln and Jefferson memorials and the Capitol Rotunda, and the majestic White House housing a good man.

During those years, I became an Eagle Scout, and was an honors student growing up outside of Boston. My patriotism and pride were unswerving, especially when Sputnik went up in 1957. To answer the call to the space race, I enrolled with half the freshman class as a physics major at Williams College, aspiring to become a space scientist and astronaut.

2. *The 1960s*. This optimistic but turbulent decade began with the Camelot years of John F. Kennedy, but was later rudely punctuated by his assassination and the escalation of the Vietnam War. Yet I clung to my optimistic call to expand our destiny into outer space. In June 1961, JFK made his historic speech to joint Congress: "I believe that, by the end of the decade, we should land a man on the Moon…as a nation we shall not floundah in the backwash…"

The timing was perfect for me. I was just graduating from college in June of 1961, and so moved to Washington to work with NASA, which was just created by the JFK vision and growing fast. I got on the ground floor of a wonderful, expansive enterprise called Apollo.

These times were not without their challenges for me. The Vietnam War was heating up, and I was almost drafted, having passed my pre-induction physical a mere week before showing up at Fort Dix for basic training. But enrolling as an astronomy graduate student at Georgetown University, I pleaded with my draft board that my role as a professional space scientist was more valuable to the nation than being in the infantry. It worked: in an unprecedented move, the draft board gave me a student deferment.

As an Irish-American, I sometimes led the wild life, organizing whiskey sour parties at cherry blossom season sunrise and throwing empty beer cans into a dry bird bath three stories down from a bachelor pad some of us guys shared. I then wrote a play satirizing the Georgetown astronomy department. A fellow student working for the CIA, true to his colors, leaked the script to the department chairman, whereupon I was thrown out with a Masters degree as consolation. Some of us donned black robes and tall conical wizard hats with the moon, Saturn and stars during our comprehensive exams. Some of this irreverence would follow me for the rest of my life when confronting corrupt systems.

Needing a job, I taught math for a few months during 1963 and

1964 at Washington's Calvin Coolidge High School learning about the realities of a half-black, half-white neighborhood in a rough sector of the city. This required commuting across town every morning rush hour, preparing lessons and sipping coffee at the many red lights on the way, and exercising the credo that the teacher must always be a small step ahead of the students.

There was an optimistic spirit then of Camelot, for example, when I met Robert F. Kennedy at a Thanksgiving dinner in 1962, where he laughed hysterically at a recent record release of Vaughn Meader's parody of the Kennedys. They were human, and they could laugh at themselves.

Then in November 1963, JFK was assassinated, which sent shivers and shocks across the nation and most of all Washington. I could walk down the block from my apartment to see the solemn procession of his casket roll by, lending a poignant reality to the horror.

After getting married in 1964, my wife and I moved on to graduate school at U.C. Berkeley, where I completed a Ph.D. in astronomy and planetary science. In 1967, I went on to become an astronaut, the first appointed to go to Mars. My meteoric career at the ripe young age of 27 brought me back to Washington triumphantly as a potential hero, respected and admired—and a bit inflated by ego.

This was short-lived, however, and the war in Vietnam preempted my space dreams. In 1968, President Lyndon B. Johnson cancelled the Mars program I was appointed to fly in, plus the later Apollo flights. Our group of scientist-astronauts suddenly felt like we were no longer needed and we dubbed ourselves the Excess Eleven. I soon quit and retreated to become an assistant professor of astronomy and space science at Cornell University alongside the famous astronomers Carl Sagan, Tommy Gold and Frank Drake. I was once again working with NASA involved in exciting planetary missions, frequently going to NASA in Washington in a positive way, ever building my career and supporting the peaceful uses of space.

But my optimism was to be further deflated by the assassinations of RFK and Martin Luther King, our escalation in Vietnam, and the dwindling Apollo program excitement. Toward the turning of the decade, hubris and militarism were beginning to replace the optimism felt in Washington during the sixties.

3. *The 1970s.* This complex relationship with Washington then became much more intimate, intense and confrontive during the 1970s, a time of uncertainty about career and ambivalence about the nation's destiny. It began with a protest against Nixon's illegal

invasion of Cambodia. In April 1970, with two other Cornell profes-sors, we locked arms and walked between a line of bus-barricades towards the White House, risking arrest for our civil disobedience. Instead, we were invited into the White House to air our grievances to some of Nixon's staff. It was there I felt the mystique of the king's palace where you could hear a pin drop, that odd mixture of awed silence and indignation. That night, I appeared with other protesters as the lead story on CBS Evening News, and it felt good to be able to express in this way, to be part of our inevitable withdrawal from Vietnam. These well-publicized protests didn't only help end the war, they began to reveal the corruptions of a White House immortalized by the Watergate scandal.

It's sad that these kinds of expressions opposing the government are no longer possible. The imperial hubris in Washington is now leg-endary, the stakes are higher and the mainstream media are shut down from dissent. The Vietnam protest experience was a precursor, as we seek new strategies and innovations such as free energy to break the ice. Protesting wouldn't be enough. We would have to articulate a vision no politician or media pundit seems to be able to give us. Washington had been calcified now and it was time to take it back before it became too late.

Vietnam was not the only seed of my discontent. NASA was los-ing its former visionary luster and falling into bureaucratic mendaci-ty. With no goals beyond the lunar landing, it became embroiled into creating multibillion-dollar miasmas such as the space shuttle, space station, and an increasing militarism. I objected to this direction, often testifying about it to U.S. Senate committees. This did not win any friends within NASA. I moved on to teaching and researching technology assessment, energy policy and national scientific priorities as a visiting professor at the University of California Berkeley Law School and then a professor at San Francisco State University and Hampshire College.

In 1975, I was appointed senior staff consultant on energy to the late Rep. Morris Udall's Subcommittee on Energy and the Environment and as a policy advisor and speechwriter to Udall when he ran for president. I felt a morally safe haven as an emerging envi-ronmentalist opposed to the perils of nuclear power, in concert with Udall and most of the Democratic congressional majority.

But my Udall job took its toll. The other top advisors were Dexadrine-popping, power-hungry insomniacs who wanted me to work long hours too, while the candidate flew figure-8s around the country campaigning vainly against the implacable power structures of Washington, of which he was unwittingly a part. The last time spent with Udall was preparing him for a speech he gave at Boston's historic Fannuel Hall, briefing him in his motel room on the talking points as he shoveled breakfast into his mouth half asleep.

Udall's wife had had it, and so had mine. He later pulled out of the race and its crazy-making while my wife began to pull out of our marriage, and so I moved into a tiny basement apartment on Capitol Hill and hung in there at a job that stressed me so much that I ended up weekly on a psychiatrist's couch for a few months, personally feeling worn out. What I learned from this experience was that, even if you were on the ethical side of an issue, power still corrupts and those who are more ambitious and ruthless were the ones that rose to the top. It was time to leave Washington once again.

In 1976, I emerged as a physics faculty member at Princeton University, taking frequent trips to Washington with my senior colleague Gerard K. O'Neill, working with NASA to envision the peaceful settlement and industrialization of space using lunar and asteroidal materials. Alongside the civilian planetary, satellite communications and Earth resources programs, these initiatives were still segments of NASA that provided hope for the future. My relationship with Washington again renewed, albeit short-lived, as once again, the militarists took over. Ronald Reagan was elected president, and the neo-conservatives pervaded every level of activity that competed with ours. NASA, Congress and what was to become the Department of Energy had lost their visions and I once again withdrew from Washington for another several years.

On one raw March day in 1979, the renowned physicist Freeman Dyson and I pulled out of Washington's Union Station on Amtrak heading home to Princeton after a scientific meeting. Barely minutes out of the station the train suddenly jumped off the tracks as we thumped along the washboard cross-ties and gravel and started to lurch over to the side. People screamed and some had been injured during the derailing. After attending to some of the injured passengers, we all walked over to a Budweiser brewery where we awaited buses to take us back to the station for the next train. This experience once again catalyzed my deep desire to stay the heck away from Washington. When would I ever learn, I asked myself.

4. *The 1980s.* In 1982 I moved to California and began working at the premier military-and-space contractor Science Applications International Corporation (SAIC). One condition I had set for the

employment was to refuse to do military or Star Wars work but instead to seek contracts on civilian NASA business, designing human Mars missions and space stations. One again, I was commuting to Washington. Once again, the military-industrial complex became my medium, albeit a civilian one. The job was lucrative, helping to send my kids to college, but my stubborn resistance to military work led to my being laid off four years into my employment just prior to being vested in my retirement plan.

My office was replaced by a former colonel in the U.S. Air Force Space Command who passed through the revolving door of retirement to SAIC as the world expert on "post-SYOP" nuclear blast detection from space. He pulled in the millions earmarked to study the strategic options the military might take after an all-out nuclear war. What a use of taxpayer money!

I also presented at a small company briefing in Washington which SAIC gave to former secretaries of defense Robert MacNamara, Clark Clifford, Melvin Laird and Howard Brown, as well as neoconservatives Richard Perle and Paul Wolfowitz. It was a creepy experience. These powerful people, who were well-paid to attend, looked and acted darkly robotic.

Once again, I pulled out of the insanity of it all. At that point I gave up all work for the defense industry. Suddenly unemployed, I worried a lot about money at first, but eventually adjusted to a new life as a freelance author and speaker. Here is my advice to those of you considering some form of noncooperation or blowing the whistle at the risk of losing your job. Fear not, you will find other work to do which is much happier. Please, for the sake of the world, get out and join us. The truth will set you free.

In 1987, Washington beckoned me again, this time the Institute for Security and Cooperation in Space (ISCOS, now ICIS). I was appointed as an unpaid board chairman and came to Washington almost monthly for meetings at their Logan Circle headquarters. Here I witnessed FBI agents pursuing us as we voiced our objections to Ronald Reagan's Star Wars program to weaponize space. We also supported Jesse Jackson in 1988 in his California Democratic Primary campaign (which he almost won) to convert the aerospace industry to helping create new jobs in the energy, the environment and other domestic priorities. Along with ISCOS president Carol Rosin, Senator Tom Harkin, Congressman Joe Moakley and some Soviet cosmonauts, we set up legislation for joint human missions into space as an indication that former enemies can become friends.

Toward the end of the 1980s, I became a citizen-diplomat during the dawn of Soviet Glasnost, speaking at the Soviet-American Citizens Summit in Washington, going to the USSR to share with their Space Institute scientists a concept for joint missions to Mars, taking a peace cruise down the Dnieper River in the Ukraine, and traveling to China. These encounters with friendly people everywhere freed me from the illusions of geopolitics and power games.

But when George Bush senior settled into his administration, the struggle to restore and renew the republic became ever more uphill, so once again I plotted to withdraw from this crazy capital, this time I hoped for good. I had to stop this addiction. One of my final gigs was attending a Star Wars meeting with 1000 contract-hungry besuited male industrialists gathered at an Arlington, Virginia hotel down the street from the Pentagon. We listened to the late Edward Teller, a Dr. Strangelove archetype with bushy eyebrows and a Hungarian accent, the father of the hydrogen bomb. He presented his vision of laser weapons in space that could zap anybody anywhere any time on Earth.

Even though by this point I was on the outside looking in, I said to myself, this was it. The neocons were beginning to take over the government, and there seemed to be little we could do about it. I vowed never to come back to Washington, except as a neutral observer or for social visits. The permanent war, the oil economy and the obliviousness of our "leaders" to solutions had won again, and I was suffering once again from outrage fatigue.

Meanwhile I joined many others exploring the unfolding consciousness revolution that could provide us with the very practices and technologies we will need for the coming transformation. Washington was not a crucible for innovations. It became even more concretized.

5. *The 1990s, the Millennium and post-9-11.* At this point, I entered the greatest void time of my life with respect to Washington, except for a few short visits to friends. This was when I walked in front of the Senate office buildings and stepped into wet concrete which almost gripped me permanently. My awkward almost-panicked self-extraction and the desire to leave an imprint would haunt me for years to come. I began to realize that the job of restoring the republic required great risk-taking and confrontation, getting out of the old situation while embracing new creative possibilities, and having the staying power to preserve the new creation and the re-creation to form and sustain a healthy collective.

My clumsy stuckness seemed to symbolize human encounters with the corruption all around mankind. From the Reagan years to Bush 1 to Clinton to the first term of Bush 2, I refused to go there any longer for any professional or activist purpose. The collective commons was unraveling and ready to go into crisis mode. I needed to be physically detached in order to make a sane evaluation of whatever choices we might have for the future.

And this freed me to do other things, like study the potential of new energy, new science and consciousness. For example, studies showed that, for the price of a few days' fighting in Iraq or a week of profit for ExxonMobil, we could research, develop and deploy clean, cheap new energy for a sustainable future. I didn't need Washington for that, and it didn't seem to need me. Moving there could only be an impediment to progress, and so Meredith and I moved further and further away, until we settled in the Andes of Ecuador, founding Montesueños, a new retreat center for peace, sustainability, the arts and new science. It is in this kind of setting where a new republic could be founded, for we not only need to return to the wisdom of our founding fathers, as embodied in the Constitution, but we must become more ecologically aware now at a global level, throwing away old tyrannies and setting new principles which will take us through the transformations we will need to make.

My last employment before leaving the country was teaching yoga to children in a tiny public grade school. Ironically both the school and we were located in Washington... California. To get the job I had to sign an oath of allegiance to the U.S. Constitution. This oath is required for public employees. At the moment of signing, I felt proud not only because I agreed with the document, but the fact that nothing about the oath says anything about obeying the illegitimate regime now in Washington D.C. I realized also I might have to come back once again to the nation's capital to defend and protect the Constitution in other ways. Here I was, two years later, ready to stand up and do just that, in front of the imposing White House.

The world nowadays is different from that of the Founding Fathers. Not only do we have to restore the system, we must remove all weapons of mass destruction from the Earth, stop our carbon and particulate emissions, and replace war and ecological destruction with initiatives such as new energy, which promise to end our dependence on polluting energy. This seems to be an impossible order for a corrupt Washington.

Unfortunately, those in charge don't want to do that. They suppress new directions and probably violently destroyed the dreams of visionaries like those of my late colleague Dr. Eugene Mallove, editor of *Infinite Energy* Magazine. As we saw in Chapter 3, Mallove was brutally murdered in 2004, probably by those whose feel their interests were compromised by embracing the new. His bold visions were not appreciated by the vested powers, cashing in on the bonanza of their lives while pulling everybody else down. A prime example was Vice President Dick Cheney's infamous secret energy task force, made up of the friendly folks at Enron, Haliburton, Big Oil and Big Coal.

Most of all, I felt I must return to Washington to express and interact with other citizens. I could no longer run away for life. Many of us have also taken the first two steps in entering the ooze: (1) to risk confronting the powers-that-be (stepping into the wet concrete) and (2) to remove their grip on us (getting out of the old paradigm before being trapped). It was time now to make a third step, that is, to leave an impression.

I was coming back to Washington to help restore the republic, expand it to its proper global stature, and embrace new visions which could truly end war, injustice, and ecological destruction. My hope in our citizen actions has always been that scores of other citizens will "get it" and we can together build a new world while preserving the precious beauty of all of nature and our beloved and betrayed country. In Part II of this series, we'll see what happened when I came back to Washington for the fifth anniversary of September 11.

II. Journey into the Belly of the Beast

A recent encounter with our beloved and betrayed contemporary Rome and an inquiry into saving civilization from the madness of our government.

Amtrak train #111 pulled into Union Station at 8:50 am on 9/11/06. This was my first proactive mission to our nation's capital in over eighteen years. The seductive multilevel consumer layout in the station was thriving with Guccis and Puccis, sushi and scones, Starbucks and Swaroski crystal. The majestic Roman arches and columns completed an image, which, during my last active years in Washington two to three decades ago, had been just in the making, then a largely empty space housing a few ticket booths and newsstands. If you were there at night, you'd have seen a few vagrants and policemen acting out their eternal conflict.

On this journey my wife Meredith and I feared for my life. After all, I was a seasoned war protester and a new energy advocate, a thorn in the Bush agenda. I was likely to be on all sorts of lists. Many of my colleagues had been murdered and I too have had my fair share of adventures as a heretic, so this return was not without its intrigue. Would I be arrested? Would the orange alert then in effect be increased to a red alert, another "terrorist" attack or martial law? Would I be followed, mugged, poisoned, shot, bombed or nuked? These thoughts ranged from a self-reflection of a sober reality projection to one of paranoia. But it was too late, we were there. I must fulfill my mission.

Walking through the morning drizzle to the van of one of the video crews for my protest gigs, we headed off to the National Press Club, where various press conferences commemorating the fifth anniversary of 9/11 were underway.

The first one we entered was a small and animated crew of 9/11 truth-seekers, researchers who just didn't buy the official story of the 9/11 Commission Report (based on my own extensive research, I didn't buy it either). The panel of experts included former Bush administration Treasury Department official and economics professor Morgan Reynolds, former Pentagon Star Wars colonel Robert Bowman, author Jim Marrs and an enclave of young activists with black T-shirts. None of the mainstream media was to be found within 100 feet of this truth-fest, they were so bought out and tunnel-visioned and inclined to dismiss us all as conspiracy theorists (while their own acceptance of the official version is itself a wild conspiracy theory).

The latest books on the table of this countercultural gathering included David Ray Griffin's *The Christian Faith and the Truth Behind 9/11* and *9/11 and American Empire: Intellectuals Speak Out*. Both books are eye-openers. Griffin, a prolific professor emeritus of theology and philosophy at the California Claremont colleges, has assembled the most comprehensive and scholarly treatments of 9/11 to date. Griffin and other notable intellectuals have concluded that the physical, eyewitness, video, confessional, circumstantial and motivational evidence of this great tragedy clearly points to its orchestration by elements within the Bush administration as a pretext for launching the wars in Iraq and Afghanistan. They also needed an excuse to discard the Constitution and Bill of Rights in favor of creating a totalitarian superpower. Griffin described the moral implications of such an unconscionable crime committed by some of "us" and the actions we'll need to take as spiritual beings wishing to end the tyranny.

To me, even the possibility that our own government either had done it or allowed it was but another escalation of outrage within my heart on how our unelected leaders could become mass murderers and torturers, controllers of the world's oil, destroyers of the environment, thieves of the public treasury, and guardians of a corporatocracy that keeps us in fear, as they ever further consolidate their power and money. We'll return to this most important point.

My next D.C. post was in front of two American flags in the lobby of the National Press Club for a couple of interviews on alternative video. This was an interesting location because hundreds of besuited mainstream journalists filed by, some condescendingly gazing at my impromptu gig. They were headed for what they thought was the real gig: a press conference held by 9/11 Commission co-chairs Thomas Kean and Lee Hamilton, still peddling their own outrageous conspiracy theory that the whole job was done by nineteen Arab terrorists with box-cutters hijacking four planes under the direction of Osama bin Laden.

I reflected at the robotic obsequiousness of the press personalities as they filed in between me and the camera. I wanted to either reclaim or desecrate the flags behind me so abused by King George's multiple flags behind all his war-talk…and most of all, that that flag has probably become a "false flag" underlying 9/11. I felt ambivalent during the interview about the overuse and subsequent desecration of the flag by Bush-Cheney in their numerous neo-fascist appearances and wondered if that flag will go the way of Hitler's swastika. It was a miracle we weren't kicked out, although once we were evicted, just to sneak back in again. I guess my astronaut credential might have helped in this city of vanities, giving me more time on-camera.

Then it was time to hop a taxi to have a meeting with Rep. Dennis Kucinich to discuss coming legislation to support new energy development. Kucinich, one of the few visionaries in Congress, understood the importance of reallocating the weaponization of the world and space to true solutions to the global energy-environmental crisis.

My 9/11 Speech in Lafayette Park. The climax of our day in D.C. was speaking at a 9/11 truth rally in Lafayette Park right in front of the White House. About thirty others, mostly wearing black T-shirts cheered, as I delivered a speech, the text of which follows:

"My fellow citizens, I have returned to Washington to protest and petition my government for the outrageous criminal actions they have performed against we the people. I have a fifty-year history with this city having played many roles, as a graduate student, as a space scientist, as a high school math teacher, as an astronaut, as a professor of science policy, as a Vietnam War protester, as a congressional aide and speechwriter, as a citizen-diplomat, as a truth-teller about our militaristic space policies.

There is a tyrant living in La Casa Blanca (I point to the White House behind me). We must find all means possible to remove him from power. He doesn't deserve to be there. He wasn't even elected—TWICE.

Regarding 9/11 and other secret operations, we demand the truth. WE WANT THE TRUTH, WE WANT THE TRUTH, WE WANT THE TRUTH!

Are any of you around here working for the government and know there's something wrong here? Are you willing to risk your job and blow the whistle? Now that's being a true patriot!

Are any of you here working for the CIA, NSA, DIA, or FBI? Raise your hands please, don't be shy, you're welcome here. All I ask is, do you really want this job? Don't you want to experience a new freedom from the shackles of tyranny, money and power? Then join us! That's what's meant by freedom and democracy.

Our new agenda must go far beyond removing these tyrants from power and getting the truth and stopping the criminal wars and recovering our rights as citizens. We need a new agenda to create a peaceful, just and sustainable Earth. Our survival depends on it.

If some of you still believe we must respect and follow this self-styled, illegitimate 'president,' let's hear from some past real presidents:

'I hope we shall crush in its birth the aristocracy of moneyed corporations, which dare already challenge our government to a trial of strength and bid defiance to the laws of our country.'

– Thomas Jefferson

'If Tyranny and Oppression come to this land, it will be in the guise of fighting a foreign enemy.'

– James Madison

'To sin by silence when they should protest makes cowards of men.'

– Abraham Lincoln

'The truth is that liberty is not safe if the people tolerate the growth of private power to the point where it becomes stronger than that of their democratic state itself. That, in essence, is Fascism.'

                                                    – Franklin D. Roosevelt

'In the councils of Government, we must guard against the acquisition of unwarranted influence, whether sought or unsought, by the Military Industrial Complex. The potential for the disastrous rise of misplaced power exists, and will persist. We must never let the weight of this combination endanger our liberties or democratic processes. We should take nothing for granted. Only an alert and knowledgeable citizenry can compel the proper meshing of the huge industrial and military machinery of defense with our peaceful methods and goals so that security and liberty may prosper together.'

                                                    – Dwight D. Eisenhower

'To announce that there must be no criticism of the President, or that we are to stand by the President, right or wrong, is not only unpatriotic and servile, but is morally treasonable to the American public. Nothing but the truth should be spoken about him or any one else. But it is even more important to tell the truth, pleasant or unpleasant, about him than about any one else.'

                                                    – Theodore Roosevelt

'The two enemies of the people are criminals and government, so let us tie the second down with the chains of the Constitution so the second will not become the legalized version of the first.'

                                                    – Thomas Jefferson

'The only way to win World War III is to prevent it.'

                                                    – Dwight D. Eisenhower

'The issue today is the same as it has been throughout all history, whether man shall be allowed to govern himself or be ruled by a small elite.'

                                                    – Thomas Jefferson

My friends, what is happening now is the very thing these former presidents feared the most: a fascist takeover of everything dear to us, with the added crime of killing off the planet itself. We must start a peaceful revolution and transformation towards a world of sustainability and justice, and this is the time and place to announce that intention. To do any less is to sell our souls to the devil and to destroy our civilization.

So, my fellow citizens, I've come to Washington today to join you in striking old roots and planting new seeds.

We need not only to take back Washington and restore the republic and our Constitution as the law of the land.

We are not only creating a movement of movements.

We are not only encouraging whistleblowers to risk their jobs and come forward with the truth.

We are not only pressing for impeachment and removal from office the tyrants in the Bush administration.

We vow never to vote for anyone who supports war and most of the Bush agenda.

We must remember that this regime was not even elected. They don't deserve to be here. (points to the White House)

Every time the media equates "Bush" to the "U.S.", every time the media equates us citizens with what they call the "U.S." as if the controlling cabal's foreign (and domestic) policy decisions represent the will of the people, our standing in the world falls another notch. I do not acknowledge this myth and do not recognize this cabal in charge as the legitimate government of the United States. The growing realization of this truth will give us the space to organize a new, peaceful visionary agenda.

I personally have spent about half of my fifty adult years in Washington. On one of my more recent visits, while walking near a newly-cemented sidewalk in front of the Senate Office buildings, I fell into some unprotected wet cement. This taught me many valuable lessons beyond the obvious one, 'watch where you're going.' This embarrassing experience showed me which steps we might take to restore the republic:

1. Be willing to confront our fear of entrapment,
2   Be willing to extricate ourselves from rigidity, and
3. Be willing to leave a lasting imprint through politically incorrect new ideas.

There are many recent successful models for creating nonviolent positive change in our policies: citizen diplomacy, which helped begin a friendship with our former enemies in Russia; the efforts of fearless peace warriors such as Mahatma Gandhi who successfully fended off the British Empire; Martin Luther King who restored civil rights to the poor; and Nelson Mandela, who led the abolition of apartheid in South Africa through the process of Truth and Reconciliation.

We not only need to recover our own Constitution. We should also draft a manifesto for peace, sustainability and justice as a template, a declaration of interdependence to precede the drafting of worldwide constitutional system that includes the relevant features of the current Constitution and brought up to date, to deal with our contemporary tyrannies, opportunities and the restoration of power to we the people.

In this process, we need to do more than clean house. We need to both restore the original house and build new houses.

Our agenda has to be visionary.

We must include bold Apollo-type programs to develop clean energy, exploring the full range of options whether they're accepted or not by the mainstream scientists.

We must create those structures that foster peace and end war, torture and illegal detention.

We must prevent the weaponization of space and eliminate from Earth all weapons of mass destruction.

We must provide for continuing support for those displaced in moving from a war economy to a domestic peace economy.

We must find new ways to control the greed of large corporations.

We must restore and enforce the Constitution and Bill of Rights.

We must ensure honest elections and eliminate private campaign financing.

We must provide health care and environmental protection and disaster relief for all.

We must make new friends of enemies.

We must stop the illegal surveillance of citizens.

We must end huge military budgets.

We must bring the National Guard home where they're needed and keep the Army off our streets.

We must require that the budget be balanced.

We must decouple from imperialistic economic policies and institutions such as the Federal Reserve, World Trade Organization, World Bank and International Monetary Fund.

But most of all, our new agenda should require that we apologize to those who have been victims of our aggression. We should work with all nations to form a global green democratic republic to address questions that affect all of us, such as climate change, peace and international law.

We should develop revolutionary clean energy sources unfettered by disbelief and vested powers. A new world awaits us if we only awaken.

"Thank you and may peace, justice and sustainability prevail!"

Escape from the Eye of the Storm. And so after greeting some of the thirty stalwart protesters, Meredith and I retreated across Lafayette Park and hopped a taxi back to Union Station, wondering why things seemed so calm here in the center of action. Civil servants, ordinary people and traffic milled around the streets and the police force was thin. No obvious CIA agents or hidden cameras were lurking about.

On the surface, everything seemed genteel, mannerly, dressy, a reverie of the tranquility of the palace I had experienced when some of us Vietnam protesters in 1970 were invited into the White House. Somehow that deceptive quiet had now spread to all of official Washington. Business-as-usual prevailed. This time we were among thirty protesters instead of the 100,000 in 1970. No mainstream media came out for this protest. In 1970, I was interviewed on the lead story of CBS Evening News.

The city seemed to be drugged out of reality as the worst violence was being unleashed from right here upon our less fortunate fellow humans in remote places, the murder of animals and the planet itself, these criminal actions perpetuated in our name. The policymakers surrounding the Mall stretching from Foggy Bottom to Congress and the Supreme Court, from the Pentagon to the White House with their illegitimate occupants, were corrupting our souls through a massive war on the world, here at the nexus of distracted consumerism, while the revolutionaries preferred to stay home. It was unreal.

The one consolation was that all this was recorded on video. Then it occurred to me that most of the 9/11 Truth folks have retreated to their computers to haul in the latest piece of research that will nail the regime. As important as that may be, it won't change things. These people needed to join the thirty others and keep multiplying so the streets were filled and we could not be ignored. Ironically, that was why my gig was curiously "safe" yet marginalized. Often, it has been said that the opposite of love is not hate, but apathy. You could cut the apathy in Washington that day with a knife.

Underneath this genteel passiveness lay a Roman force so globally massive it sometimes felt like resisting was a lost cause, that that nuke would drop at any moment so why not whoop it up, eat drink and be merry. Go to that special French restaurant on K Street as a trough of lobbying, or at least have a sushi or scone at Union Station while awaiting the train back to New York.

Peeling the Onion of Truth. We've had a coup d'etat, after a buildup of at least decades marked by a disastrous foreign policy, greed, militarism and environmental neglect. Symbolically we can mark the precipitous decline of the American Empire at the start of this millennium, at the time Bush was illegally selected to be president in December of 2000. In January 2001 he was crowned King of the World during a lavish inauguration party hosted by his buddy Ken Lay of Enron. Soon after, Cheney met with his infamous energy task force made up mostly of oil, coal and natural gas plutocrats and their financial cronies, carving up the world for neo-conservative fascist invasions soon to come.

The neoconservatives, at this writing in power, had always craved for wars in Iraq, Afghanistan, increased military spending, deploying weapons in space, and backing out of treaties. As their published writings have clearly shown, they desperately needed a pretext. That pretext was obviously provided by 9/11.

Whoever ordered 9/11, we can say that only two elite groups benefited: a small selection of Muslim extremists, but much more, the Bush administration itself and its corporate and political cronies, who have pocketed trillions. They won this one, and even if all morality and sensibility points to their guilt, they've still won, even if they're on their own suicide mission to end their careers in dramatic sacrifice for the greater greed. Haliburton has already won. It's just too bad the rest of us have been so bilked by the extremists.

It became time to remove them from office and have a coup d'etat of the coup d'etat. One scenario was impeachment (which could have taken place with a Democratic majority in Congress during 2007 and 2008). But impeachment was "taken off the table," and most of the Democrats have been spineless and join the elite arrogance of the sole superpower, ever decaying, while most of the rest of us still alive on Earth cry out for big change.

But even that obvious step might not happen because of the notorious Republican purging of voter registration lists and rigged machines. More than three-fourths of Americans don't want to re-elect their representatives or Senators anyway. Anarchy sometimes feels better than the current scenario of control by the elite and their media mouthpieces. The economic crisis of 2008 is but the latest step to consolidate greed and corruption in the U.S. government.

But the only lasting solution will have to be a world government that will take most of the power away from the U.S., especially in militarization, economic imperialism, human rights violations and the destruction of the environment—amplified by an obsequious media. The U.S. government seems to have outlived its usefulness. We need to face that and recover our beloved Constitution to help get us through this transition to a global green democracy. After all, the whole purpose of a democracy is for the people to choose a government they want, not the one the very rich want.

Think about it: when the going gets rough and the Democrats are compliant, and in the absence of a viable third party, one of the best solutions for replacing a corrupt government might to create a parallel government (this happened in Mexico when people flocked to the streets after a likely rigged election). Some former corrupt Ecuadorian presidents were peacefully ousted by demonstrations in the streets. Why not in America? We should be doing that now, and we need for the whole world to join in, a world government of confederated states, just as the U.S. had done for its own states. We need a process of truth, reconciliation and a global green democracy charged with jurisdictions that affect all of us.

Arriving at the truth and taking the necessary actions is like peeling an onion. The outer layers are the roughest, least aromatic and least edible. They might represent the policies of the Bush administration, very crude, very dispensable.

The next layer is a bit more edible yet is also crude and politically correct and warmly embraced by the corporate Democrats and liberal portions of the media. Some of these pundits may decry the Bush policies but don't really want to make the necessary changes. Barack Obama, Joe Biden, Hillary Clinton and John Kerry come to mind.

Peel the onion some more and we come to the articulate expressions of an Al Gore who eloquently points out the problems of global warming and climate change. Or those who feel that we should withdraw from Iraq after all, such as the expressions of a Ron Paul, John Murta or Dennis Kucinich. This is just barely tolerated by some of the elitist mainstream. Just barely, yet not enough to upset things too much to sabotage re-election. This takes political courage but hardly scratches the surface.

For the peeling of the onion has just begun. As we peel to lower layers, truth becomes more difficult to embrace, because we begin to enter the realms of political incorrectness. The gap between truth and image here is greater. The journey reaches new levels: did the Bush administration really orchestrate 9/11? Should we not impeach and convict the neo-cons and robber barons of Wall St. and shut down the entire apparatus and start anew? Should we send them to jail for violating our rights for starting an illegal, immoral war and for their raids on nature, the people of the world and the public treasury? Should we not cut back on our carbon-based fuels and rejoin the world community by embracing the Kyoto protocols and various treaties trashed by the Bush administration? These questions seem to occupy the next layer.

The next layers may make us cry. What painful steps will America need to take to relinquish its global economic and military hegemony, to revision its foreign policy and return to caring for its own citizens and nature? What sacrifices must the elite and some of the rest of us make to stop the imperialism and confess these sins to the rest of the world? How can we rebalance our budgets and eliminate our economic and military fascism?

The next layer we come to is barely perceived; it's just too much to consider in any normal discourse, yet it may provide the ultimate solution to the most basic problem we face: that we are killing ourselves and all of nature. We have solutions, the most pervasive of which is new energy, the prospect of a truly clean, cheap, decentralized energy economy that would totally end the Bush nightmare, the Republicratic Congressional nightmare. I am not talking about the conventional renewables such as solar, wind, biofuels and hydrogen fuel cells. But this layer is not even discussed, because it isn't believed. This is the layer some of us are working on behind the scenes.

My book *Re-Inheriting the Earth,* my postings on www.brianoleary.com and Wade Frazier's writings on www.ahealedplanet.net, Tom Bearden's www.cheniere.org, and Jeane Manning's and Joel Garbon's new book *Breakthrough Power,* describe the promise of new energy. We need to get beyond the denials of mainstream scientists who really don't know what they're talking about. I have many reasons for asserting that. The so-called laws of physics and the limited visions of politicians and media are forever being broken by new visions. I can say that with the authority of physics professorships and an impeccable publication record for 45 years. That is my piece of the puzzle, and we need to get to that layer before it's too late.

Are we ready? We'd better be. We don't have much time for more denials and fears and posturing.

Meanwhile, I recommend you have a good cry while you keep peeling the onion of truth, and to get over the grief of change while you embrace a new world we can still have.

III. Towards a New Energy Truth Movement
and Other Truth Movements

After our return from the belly of the beast, the forces of polarization and violence seem pitted against a small but growing movement of movements to seek the truth of so many things smothered in lies.

One week after our trip to D.C., Meredith and I flew out of JFK airport back to our Andean retreat. As we made our way across Manhattan in the early hours among thrashing helicopters paving the way for Mr. Bush's arrival for delivering yet another mendacious speech to the U.N. On the previous day, Venezuelan president Hugo Chavez had given a fiery speech, accusing the Bush administration of embodying the very evil they claim to be fighting. And Al Gore had just given a strong (for him) speech at NYU law school sending out the alarm of coming episodes of catastrophic climate change and calling for a freeze on the growth of carbon emissions.

Our flight back home had only 20 passengers, so that my own carbon emissions quota in taking that flight would be for more than a lifetime, and I shouldn't travel again. We can all reflect on how we contribute to this disaster that the oligarchy has decided to foist upon us, simply because of an idea advanced more than a century ago that we should depend on the incomplete combustion of hydrocarbons found in the Earth as the mainstay of our energy use. The great wealth generated for the few, the convenience of using these fuels, has so swept over us we seem to be locked into a paradigm sure to send us all to disaster, what Gore calls an "inconvenient truth." We are so distracted either collecting our money for evil deeds or fighting those deeds so we can't get to our real agenda: creating peace, sustainability and international justice.

But it was not until we got back for a few days that the clincher came in. The weak-kneed, obsequious U.S. Congress just passed the Military Commissions Act, perhaps the most draconian measure against its people in the entire history of the U.S. and perhaps since the enactment of the Magna Carta 800 years ago. The very founding principles of our Republic and Constitution were destroyed in one vote to deny *habeas corpus*, the legal right of a defendant to face his accuser in a fair and speedy trial. This new law of the land gave Mr. Bush and his minions the authority to capture, detain, torture and murder anyone in the world, even retroactively. It will take some time for the Supreme Court to declare this atrocious legislation unconstitution-

al. By then it might be too late, because the court could become more
packed with more neoconservatives or we'll have more true or false
flag terrorist threats or attacks to trigger a red alert and martial law.

How could we talk of a world government if the most powerful
government on the planet, *my* government, keeps moving away from
the peace, sustainability and justice necessary for our own survival
and that of nature? The New World Order agenda embraced by the
rich and powerful is not only infecting the U.S., it is growing into
such new entities as the North American Security and Prosperity
Partnership (SPP), another "globalization" project. The SPP promises to complement the World Bank, World Trade Organization and
International Monetary Fund to line the pockets of the rich and to
spread the American Empire like a totalitarian rash, combining the
forces of neoliberalism and neoconservatism. One of the main speakers at the flagship meeting of this group was then-U.S. Secretary of
Defense Donald Rumsfeld. He and other speakers extolled the virtues
of the megacorporate vision of "free trade," military control, secrecy,
surveillance and energy policies that lock in the perpetuation of burning hydrocarbons.

Mary-Sue Haliburton, in a 2006 post, called that the "Traditional
combustion-energy paradigm at high-level negotiations under North
American SPP. Scheduled to begin to exercise power authority by
2007, the SPP will place three nations in the continent under 'harmo-
nized' laws and a unified administration. If that is not stopped—and
we appear to be past the tipping point—will any of us recognize our
society? And will it still be possible to shift the energy paradigm
under such a political paradigm shift?"

I say all these things with a sense of remorse that the world's leading nation of innovation has sunk so low and prevented us all from
creating the abundance we really deserve.

Government of, by and for the rich. When we look at the increasing consolidation of power vested in Washington, it seems to put a
damper on even talking about a global green republic or democracy
that would appear to be the polar opposite to what is actually happening. The realpolitik of the imperial agenda makes discussion of any
form of regional or world government moot. Anarchy would be preferred to such a fascist direction, moving ever further away from what
we need to survive. Yet we should hold the vision anyhow, because
the void created by the self-destruction of the empire will come and
will give us opportunities to enact the new. I'm still hopeful.

The 18-minute video of my 9-11-06 speech in front of the White House was soon posted on Google, and it sent several unexpected ripples back to me. I guess you could say it was my fifteen minutes of fame in this century, replete with political incorrectness and unpolishedness and attended by a motley but dedicated crew of about thirty 9-11 truth-seekers. If anyone were to want to dismiss me as a "conspiracy theorist" all they'd have to do is quote me out of context, see the scene, and I am marginalized for life, all credibility and credentials lost. In fact, some of my closest friends were concerned that I might have blown my image, and that my effectiveness as a new energy activist has been horribly diluted by this performance. Maybe.

Things are different than they were during my earlier years in Washington. How was it that in 1970, as a young professor from Cornell, I joined 100,000 other peace protestors to stop the invasion and war in Vietnam? Then we were invited into the White House and appeared that night on network news. My honor and reputation were not at all tarnished in that action.

In this century, appearances and soap operas seem to carry the day, especially in politics. At this writing, we see that the "news" still dishes out a barrage of the administration-this and that as a drama unfolding with these people on center stage while the rest of us are regarded as disenfranchised spectators. All of Bush's appearances have been on Hitlerian right-wing stages adorned by flags and expensive sets paid by taxpayer money, ranging from the Top Gun "Mission Accomplished" fiasco on the carrier deck announcing the end of major combat operations in Iraq in May 2003 to the illuminated cathedral in New Orleans where just down the street black people were dying from his neglect of helping the victims of Hurricane Katrina. The Reagan presidency brought in the Actor with a capital "A" and now all of official Washington, especially the White House itself, is one big expensive stage set and photo-op to aggrandize the King and his attendants.

The movements of this century have fragmented into ineffective cul-de-sacs, each one vying for its place in the sun, each one failing to make much of a dent, each one marginalizing the others even further out of fear of losing credibility. The peace movement, the environmental movement, the human rights movement and the fiscal responsibility movement are now more often dismissed by the mainstream as unpatriotic conspiracy nuts or radicals aiding and abetting the terrorists.

The movements aren't talking to each other. So my 1960s activist colleague and friend Keith Lampe, alias president USA-in-exile, suggested that we need now a movement of movements under the banner of seeking the truth. So why not use the 9-11 Truth Movement as our launch pad? This inspired me to come to Washington to do this gutsy gig. If some people, even friends, see this as a dilution of my credibility, so be it. But I don't have to please anyone in American political circles, to build a career any longer in their midst, or appear to be professional. That's the freedom of being "retired."

But I do pay a price for this. Addressing the truth can be sacrificed on the altar of image. In Washington, a media consultant came up to me with her business cards to the effect that I needed to improve my image and could use some coaching. The media join in with its own spin about *how* things should be done, not *what* should be done.

And so we have a battle going on here of truth versus political correctness, based on the fear of being marginalized, especially when we try to cross-pollinate movements. What would happen to the new energy truth movement if one of its principal proponents is also an advocate of the 9-11 truth movement, whereas another new energy person has already made up his mind that the official 9-11 story is correct and the matter is closed? Many constituencies overlap, mesh, and bob in and out of truth-seeking. It's a messy business. But we need to combine the truth movements nevertheless. Is it a threat to anyone in one truth movement to seek the truth of another matter? Of course not. Unless you're guilty about something, why not go for *all* the truth? The only thing stopping us is our own illusion of risking inquiry across the board. Image and political correctness have overwhelmed us so far in this century.

Yet there are also more serious risks in seeking the truth, for sure, and this might be the nub of a second reason why people may not want to be associated with me and my renewed "radicalism." They are afraid for their own lives, and I should be afraid of mine too. If our Constitution and Bill of Rights are in fact abrogated by the forces of tyranny, then we face our own personal struggle of how far do we go to protest that tyranny before they send out a contract for me or for any of my associates to be detained or eliminated?

The fear of death and torture is the second reason for others to distance from us truth-seekers, because we are still small in numbers and therefore easily marginalized. It's ironic that the very protection we need, strength in numbers, is the very thing we don't get simply because people are afraid. Abe Lincoln said, "To sin by silence when they should protest makes cowards of men."

To help stoke this fear and to stoke the desire to distance oneself from the protest demonstrations, you'll notice something eerie about seven minutes into the tape of my speech. Just as I was whipping up some chants for the truth at this rather tame public gathering, Dave Shaw, the cameraman who was recording the speech, focused on two men walking across the roof of Fortress White House with what looked like rifles pointed at me. Snipers! While it does call attention to the fact we all need to "cover our asses" and we also need to come out in greater numbers, it shouldn't discourage us. Said Ben Franklin, another patriot I love, "They that can give up essential liberty to obtain a little temporary safety deserve neither liberty nor safety."

So those who stay at home and are wary of the image created by marginalized protest, I ask you to restore your patriotism and hit the streets. Join us! It's amazing how some critics of my speech felt I advocated the overthrow of the government. Nothing could be more untrue. As recently as five years ago, I swore to obey and honor the Constitution and I still do. I merely expressed that our current government is illegitimate, corrupt and controlled by treasonous tyrants. We need to impeach, convict and move onto brighter agendas. Again Jefferson: "The two enemies of the people are criminals and government, so let us tie the second down with the chains of the Constitution and the second will not become the legalized version of the first."

To me, the most unpatriotic act is the Orwellian Patriot Act itself, the Military Commissions Act and the shenanigans of our administrations. It's more patriotic to hit the streets demanding the truth, even if the venue is off-the-wall and unpolished and not picked up by the media or recognized by mainstream cultural institutions. That doesn't matter. What matters is just showing up, which is 80% of life, so says Woody Allen.

Again, the most patriotic act we can perform is to seek the truth of everything the government does. This process is what abolished apartheid in South Africa, called Truth and Reconciliation. May this be a model for our country in these perilous times. And that's why I showed up in Washington that day. To me, the 9/11 Truth Movement is a template for what we can do across the board so that some day, thousands, then millions, will go to the streets, lock arms and march for the truth—even in the presence of illegal police riots.

I can't say who did 9/11. But what I can say is that, as a scholar, I have researched the events of that day, and there is sufficient reason to believe that the case should be re-opened. The perpetrators of this monstrous crime need to be brought to justice. It is ludicrous for the Bush administration and his congressional lackeys to use this solemn occasion as a pretext to wage illegal wars for oil and domination, to cause the deaths of even more innocents than the victims of 9/11 itself while opening the door to more terrorism than ever. If 9/11 itself has been so abused by our leaders to commit acts of aggression, so then 9/11 truth can also be its counterpoint to bring to justice those terrorists who really did it, and those who allowed it.

But we shouldn't stop with demanding 9/11 truth. We also need to fearlessly disclose and convict the architects of electoral fraud truth, foreign policy truth, Pentagon corruption truth, fiscal truth, torture truth, surveillance truth, new energy truth, and you can add to this to make a long list. That the president, vice president and Congressional leaders are liars is now well-established, and their protection of executive secrets is legendary. Do we not have the courage to face the truth in all sectors of our society and to fire our leaders who have so betrayed us? For only with facing the truth will we be able to reconcile our society. Our movement of movements must first be focused on the truth, and we are still searching for an orverarching theme to describe and package all of this.

In 2004, we visited South Africa and I gave some lectures there on new energy truth. The reception filled our hearts with joy. Because that country had one decade earlier gone through the gut-wrenching yet healing process of Truth and Reconciliation, they could see through the falsehoods of U.S. imperialism. They were open to seek new truths.

And there is no greater truth than that of free energy. I believe it will save the planet. Yet, the suppression of its promise for 100 years is so well-documented for those who seek that truth.

IV. Second American Revolution or Global Transformation?
A 2008 Retrospective:

Call for a return to public morality and fiscal sanity, reported from two cradles of liberty

"Banking institutions are more dangerous to our liberties than standing armies...the principle of spending money to be paid by posterity...is but swindling futurity on a large scale...The issuing power (of money) should be taken from the banks and restored to the people to whom it properly belongs."

– Thomas Jefferson

"Whenever a Great Bipartisan Consensus is announced, and a compliant media assures everyone that the wondrous actions of our wise leaders are being taken for our own good, you can know with absolute certainty that disaster is about to strike...The bailout package that is about to be rammed down Congress' throat is not just economically foolish.   It is downright sinister.   It makes a mockery of our Constitution, which our leaders should never again bother pretending it is still in effect. It promises the American people a never-ending nightmare of ever-greater debt liabilities they will have to shoulder."
                                          – U.S. Representative Ron Paul,
      "The Creation of the Second Great Depression," Sept. 25, 2008

"Game over. The Federal Reserve itself is in danger. So, it's on to Plan B: which is to dump all the toxic sludge on the taxpayer before he realizes that the whole system is cratering and his life is about to change forever. It's called the (U.S. Treasury Secretary) Paulson Plan, a $700 billion boondoggle which has already been disparaged by every economist of merit in the country...the bailout does not even address the core issues which have been obscured by demagoguery and threats. The worthless assets must be written-down, insolvent banks must be allowed to go bust, and the crooks and criminals who engineered this financial blitz on the nation's coffers must be held to account."
                         – Mike Whitney, www.informationclearinghouse.info,
                                                              Sept. 27, 2008

"Henry Paulson cashed out at Goldman Sachs in 2006 at a half-a-billion dollars. And now he goes to Washington to bail out his buddies...King George IV (is) at work again. (This bill) has no comprehensive regulation and all the other changes to make Wall Street accountable, instead it allows Wall Street to create a corporate state or what Franklin Delano Roosevelt called fascism, which is government controlled by private economic power, represented by people like Henry Paulson...there's nothing in this bill for homeowners. There's everything in this bill to bail out the bankers who actually created this problem with those out-of-control speculative financial instruments. Once this (bailout) happens, it's not going to be reversible."
                               – Ralph Nader, independent candidate for president,
                                              Democracy Now! Sept. 25, 2008

"No economy can grow at steady exponential rates; only debts can multiply in that way. That is why Mr. Paulson's $700 billion give-away to his Wall Street colleagues cannot work. What it can do is provide a one-time transfer of wealth to insiders who have been play-ing the debt-credit system and siphoning off its predatory financial proceeds to themselves (and) to take Mr. Paulson's $700 billion and run...It's a giveaway, not a bailout. A bailout is designed to keep the boat afloat. But the existing Wall Street boat crafted by the invest-ment bankers seeking to unload their junk must sink. The question as it sinks is simply who will be able to grab the lifeboats, and who drowns...the sellers (of this junk to the government) will take their money and move it abroad to a "hard" currency country. This will help collapse the dollar. Up will go gasoline costs and prices for other imports. America will be turned into a Russian-style post-Soviet economy, having endowed a new domestic kleptocracy of insiders."

– Dr. Michael Hudson, economist, www.globalresearch.ca,

Sept. 24, 2008

I write this on September 11, 2008, the seventh anniversary of 9/11. While the truth-seekers continue to dig and present ever more evi-dence of criminal neglect by the U.S. government, the decline of the Empire is ever more evident. Both the Democratic and Republican conventions featured unprecedented police brutality directed toward peaceful protesters and alternative media *even before they took to the streets.* Illegal violations of the civil rights of innocent civilians have transformed America into a totalitarian police state. Something will have to give. Earlier this year, I had been invited to speak at a con-ference in Washington, D.C. on April 19, 2008, the 233rd anniversary of the famous ride of Paul Revere. I was asked to speak about the UFO coverup, following an appearance on the History Channel, when I said that NASA and the military were hiding evidence of the UFO/ET phenomenon and some of the technologies that are being back-engineered such as anti-gravity propulsion and free energy.

The conference was billed as "The Insiders." I guess that, even though I hadn't been an insider for decades, I would at least be able to give my views as an outsider. But, much to my surprise, about one month before the conference, some individuals in the government who were specifically part of the cover-up and disinformation apparatus of the U.S. military were added to the program.

I had mixed feelings about doing this: should I risk going out in flames, just go there and declare that the emperor wears no clothes? Why was the speakers' program now including members of the infamous "Aviary," government disinformation operatives charged with *covering up* UFO/ET evidence? Why were these spies so blantantly in our midst? What were *they* doing in a truth-telling session? In the end, I cancelled the trip, realizing that the Washington environment has become so twisted that ordinary people were in constant fear and awe of the most powerful, of the deceptions that keep happening there. The September 2008 fiscal crisis is but the latest of the crimes emanating from the cesspool.

My last trip to the U.S. was in October of 2007, when we attended my fiftieth high school reunion near Boston. As I walked the cobblestone streets of Boston and Bunker Hill in Charlestown, I began to reflect on what it must have been like back in 1775, when a few courageous souls stood up to the power of the British crown. Must we do the same now? Or could we be more enlightened about how we can do this without firing a single shot, without lowering ourselves to the methods of tyranny?

What was it about the American Revolution that sprang from 140 years of colonizing by British expatriates, whose beliefs were not always that "pure" within themselves? As I strode along the beaches of Plymouth, I gazed at the tiny *Mayflower* and pondered about the great courage and motivation of the first Pilgrims. But I also wondered about the slaughter of the Indians and the prejudices against those of different beliefs that the colonists simply couldn't handle. The American Revolution didn't end the violence that continues to this very day almost four centuries after the Pilgrims' first landing. What should we do? How could we lift ourselves out of our cruelties? This, of course, is the tragic human condition that now America seems to have both the responsibility but also the power to change. It is a sober recognition that the origins of American culture can be rooted in the violent heritage of the Anglo-Saxons (e.g., see the excellent 2008 book by George Williston, *This Tribe of Mine*).

I reflected that there come times in the affairs of humanity that call for the strongest and swiftest possible action to remove a violent dictatorship, and to redirect our energies toward a just, peaceful and sustainable future. Since the Constitutional redress of grievances necessarily involves a first step of impeachment and prosecution for "high crimes and misdemeanors," and Congress is not up for the task the Constitution requires of them, we will have to find other methods of shifting the power structures of my native country. This will require concerted action by a critical mass of many courageous citizens.

Such a time is not only overdue, it must be captured soon for our own survival and that of the natural world. It may be too late already for able and compassionate Americans to take the lead, but it's surely worth a try. It is our moral obligation to do so, at the very least, as reparation for having ravaged the planet for so long.

The world is waiting, but some of the rest of humanity is not going to wait for long. They are creating their own crucibles for change. In the long run, no superpower is going to stop that. New democracies are being born; Meredith and I happen to live in one now emerging— Ecuador.

If you are not yet convinced of the seriousness of our current dilemma, I refer you to the many unfolding critiques by academics, progressives, conservatives, historians, libertarians, economists, foreign well-wishers and religious leaders of all stripes, on how our own government and their corporate accomplices continue to be at war with the rest of humanity and all of nature. These facts and opinions flood in from all over the Internet and foreign press, but remain unacknowledged by a corrupted U.S. media.

Whether it's the Iraq war, the potential for an Iran war, the torture of innocent victims, the abrogation of treaties and human rights, the squandering of public funds, fraudulent elections, the inaction towards reversing environmental destruction or the secrecy and corruption, the story is the same: a grievous abuse of power and a violation of the public trust have taken place.

In Pentagon Papers whistleblower Dr. Daniel Ellsberg's recent words, "a coup has occurred" and will only become worse if we don't take action to remove those now in charge of our collective destiny. The growing consensus is that the mainstream Democrats and media have caved into the pressures of war and profit and have become complicit in the crimes. We can't count on them. In fact, they will probably have to go too, along with all the lobbyists and bribers and private campaign donors that have so separated public policy from the will of the people.

I write this on the eve of an election in which a fairly-elected President Obama, itself an uncertainty, would have to do an about-face on his Wall Street and militarist supporters and still survive his term(s) to make the kind of difference we need. Is this possible? Not likely. To wait for an improbably reformed executive, while a refreshing possibility, does not address the *systemic* changes we will need to take to have any chance to turn around the biocide, disaster capitalism and suppression of an energy solution revolution.

The First American Revolution Started in Boston. Over 230 years ago, against all odds, a small minority of our American ancestors courageously succeeded in resisting and ejecting the mighty British Empire. It all began right here, in and around the Cradle of Liberty—Boston, Massachusetts. But what's unprecedented in these times is that there is little room for those of us left on our poisoned planet to equivocate, to rationalize our continuing support for immoral leaders under whom we have allowed ourselves to become impotent subjects of fascist rule. So how can we do this? That is the burning question for us all to ponder.

Make no mistake; humanity and all of nature are plunging into a Dark Age by the actions and neglects of the most powerful individuals, governments and corporations on Earth, who really don't care about our future. We see the warning signs all around us. But this call to action needs to happen a lot faster than the 150-plus years it took for the first colonists who landed in Plymouth (1620) and the revolution and founding of the American state (1775-76). To succeed, the new revolution will have to take place within the next few years, whether it begins in our country or elsewhere in the world.

It is time to once again to dump the tea of tyranny into our troubled waters and restore the rule of law so eloquently promulgated by our forefathers. Isn't it ironic that "the land of the free" has so rapidly become "the land of the foe," comprised of ruling sociopaths and their plundering cronies that we have allowed to terrorize us while we slept?

I think our current condition is like being on a *Titanic* captained by a power-crazed skipper and his minions, a supine Congressional crew, an aristocracy seeking the few available lifeboats, and the rest of us as supplicant and frightened steerage passengers. The ship is listing, soon to be stern-up, and then sink into a dark and icy abyss of nuclear war or an unlivable environment.

Any baby born nowadays joins the Earth herself in this joint struggle against a totalitarian, genocidal and biocidal nightmare that has visited our land and invaded many others for so many years now.

I was also born in this city almost 69 years ago into the tradition of revolutionary spirit of social change. Throughout history, many true patriots have come from this part of the world. By using the term patriot, I am not talking about the New England Patriots football team or the deceptive and Orwellian-named Patriot Act or some neoconservative fantasies of conquest or right-wing militia rebellions. The term, like so many others, has been co-opted and we need to restore its true context.

The spirit of this birthplace of liberty must now join the cradle of Earth in crying out to heal our collective wounds and embrace solutions to the human condition. Adrift in a blood-soaked ocean of insanity, we must once again unite and arise to our empowerment. May the richness of our revolutionary heritage help inspire what we must do now.

Here are just a few examples of some true Boston-area patriots throughout our history:

Paul Revere and Sam Adams for their spark and spunk in starting the Revolution;

Joseph Warren and William Prescott for their courage in battle;

Henry David Thoreau and Prof. E.O. Wilson for their deep respect, understanding and eloquent expression about the wonders of nature and negative human impact;

Ralph Waldo Emerson and Prof. George Wald for their passion about civility and peace;

Mary Baker Eddy for her courageous stand on the healing power of the mind; and

President John F. Kennedy for leading the human spirit towards the Moon as an inspiring group achievement;

Even up to the current moment, Boston is still supplying pioneers in the finest revolutionary spirit:

Professors Noam Chomsky and Howard Zinn for their intellectual rigor and honesty in their critiques of American hegemony;

Dr. Peter Hagelstein and the late Dr. Eugene Mallove of MIT for their courageous work on breakthrough clean energy research which has been ridiculed by powerful interests, even among some of their own MIT colleagues; and

The late Harvard professors John Mack and Timothy Leary who risked their careers to explore the inner workings of the mind and contact with nonhuman intelligence.

Countless other brave thinkers come from Boston.

A Second American Revolution? The first small numbers of us who have the courage to stand and be counted must do so with a fervor unfelt for more than two centuries. These actions might expose us to the same kinds of risks taken by the handful of brave signers of the Declaration of Independence. About half of them were later murdered, prosecuted or had their property confiscated. These actions are not for the faint of heart. Some of you might see this expression redundantly polemic and unprofessional. Still, I urge each of us to examine the evidence, no matter how damning, of those truths that will help free and lift us out of our malaise. Only then can we regain control of our paths into the future.

Personally, I have read thousands of articulate reports in the literature as prima facie evidence of repeated crimes our leaders have committed against the rule of law and the rights, indeed the very existence, of untold human beings throughout the world. I am sure many of you have also looked at the evidence. Yet I am stupefied and outraged at the continuing degree of corruption and destruction unleashed by those at the top, covered up by lies and immorality that go way beyond the pale.

9/11 Truth as a Pretext for the Second American Revolution. A number of truth movements are gaining momentum now, including the undoing of our foremost "sacred myth" or "official story" that 9/11 was carried out by Arab terrorists who "hate our freedoms" and who must be defeated in an unending global crusade. What nonsense! Under the veneer of this propaganda lies our denied pretext to invade the Middle East for its oil and for a permanent military foothold. Meanwhile, this entity called the "U.S." continues to commit genocide and ecocide wherever and whenever the president desires. The 9/11 disaster supplies the Empire its long sought-after power-grab. It also provides a motive to execute or allow the 9/11 attacks themselves.

The overwhelming majority of those of us who have taken the time to examine the evidence, conclude that the official story of 9/11 is a pack of lies and is an unfounded conspiracy theory in and of itself. (1) While many of us are not ready to hear this, a true patriot will. Regardless of the consequences, the truth must be confronted and not avoided.

Only when we have the courage to remove ourselves from our paradigm paralysis and muster the discipline to examine the evidence for what it is. Only when we can uncover the deeper truth of the transgressions of those whom we had entrusted to lead. Our revolutionary task is to suspend disbelief that 9/11 and resulting reign of fear may have been a cynical inside job perpetrated by the Bush administration as a pretext for preplanned wars against Afghanistan, Iraq, and Iran. At this writing seven years later, America seems to be undergoing a second 9/11: the engineered transfer of ever more wealth from working people to the financial oligarchs in charge. How could free energy ever be introduced in such a crazy world? We'll need to get away from such madness to be able to incubate and introduce the new energy technologies far, far away from the controllers—if that's possible. They don't at all seem to have the best interests of anything in our creation except for themselves.

Uncovering the truth of 9/11 requires the utmost attention from ordinary citizens who are just now beginning to understand the pattern of motives and evidence that clearly point to treasonous actions at the top. Truth-telling precedes revolution and reconciliation. But it would be a grave error to await for the final verdict on who did 9/11. We don't need a 9/11 to impeach, indict and incarcerate our genocidal leaders and to replace them with those who will peacefully lead us through the brunt of the Second American Revolution.

It's interesting that the planes that struck the Twin Towers on that tragic and bloody morning took off from Boston. Perhaps history will show that our new millennium actually opened with a dramatic show of tyranny reminiscent of the time Bostonians were warned "the redcoats are coming!" Isn't it ironic that the Second American Revolution may have also begun in Boston? Perhaps we have gone full circle here, except now the tyrants come from our own government.

Yet the possibility that the 9/11 attacks were part of a massive false flag operation is so egregious, we are surely ready for another Boston Tea Party, a call to arms against the greatest tyranny of our times. The supreme crimes of 9/11 and the resulting illegal and immoral invasions in the Middle East, the environment, the public treasury and our freedoms should cause us all the greatest possible concern. Beneath these acts lies the malicious intention that the Bush-Cheney warriors have taken us far from our true moral purpose: to restore ecological balance to Earth. Wrote Prof. David Ray Griffin at the Claremont School of Theology:

"The violence of 9/11, along with the official narrative thereof, distracted our primary attention away from the relation between humanity and nature and forced it back to human-vs.-human issues." [1]

We must now go beyond the specifics of uncovering the self-evident truths of what has befallen us. This is more than an academic, economic, political and legal matter that compels us to change course. Make no mistake: we are in a global transformation here, not a reformation. I am going to suggest that our future planning must go beyond articulating the moral basis of removing our leaders and restoring the laws of the land. We must now consider embracing real solutions, free of institutional vested interests. This needs to be a concomitant effort of undoing the old and creating the new, directed towards restoring public sanity, justice, peace and a clean environment.

Restoring the U.S. Constitution as our Guide. To help guide us through all our confusion and apathy, we are fortunate enough to have a structure and compass to navigate forward. Gratefully, the U.S. Constitution and Bill of Rights can provide the framework from which to launch our futures. Our compass must be a kind of ethical morality that I define as the ability to discern right from wrong, to follow the Golden Rule which states, "Do unto others as you would have them do unto you." Not "He who has the gold rules." I believe that we can begin to restore proper morality by implementing the rule of law, as spelled out in the Constitution. The first logical step is impeachment, and if that won't work, other means will have to be found.

Beyond our own sense of discerning what is right and what is wrong, our actions to follow will take a new kind of raw courage to restore our democracy in the face of adversity. We have no other choice. Whether we impeach these multitudes of corrupt leaders, or demand their resignations or go to Washington and bang pots until the tyrants leave, or compel the military and police to disobey illegal orders to attack Iran [2] and our own citizens, or ask government employees to blow the whistle for unconstitutional executive actions, or call a general strike and massive boycotts and sit-ins, or establish parallel governments, or create a temporary anarchy, something must be done to evict the traitors, using actions analogous to the Boston Tea Party. Just as importantly, we need to become free to begin to embrace the real solutions to our dilemma, solutions that wait in the wings for their cultural opportunities such as an energy solution revolution.

The first steps in the Second American Revolution, therefore, are to: (1) remove from power the self-designated "deciders", who are supposed to be working for us and not the other way around, (2) restore the rule of law, the Constitution, Bill of Rights, Nuremberg and Geneva conventions, treaties, and privacy and habeas corpus rights, and (3) perform some massive intention experiments described in the next chapter to make that so. Whether you're on the Left or Right, Republican or Democrat, Libertarian or Independent, we're all in this sinking ship together, created mostly by the stupidity and greed of a tyrannical elite. It is time to take action against these policies and create new ones.

Those individuals holding the most powerful positions in the land all took a solemn oath to uphold and defend the Constitution of the United States. Most all of them have utterly violated that oath, and, as a result, they must be removed from their positions.

On repeated occasions, I've taken my own oath very seriously, as a NASA employee, as a scientist-astronaut, as a Congressional adviser, and as a grade school teacher. On each occasion, I studied these laws and agreed with them. Taking the oath was a very patriotic and uplifting experience for me each time.

But there's nothing in the oath about swearing our allegiance to our leaders. If that were required, my answer would have been a definite, "NO!" Fortunately, the wisdom of our Founding Fathers relieved us from such a burden, although many people do not realize that. The Founders would have detested the many current violations our so-called leaders have made against our First Amendment rights to speak out. These individuals at the top are the ones who have abrogated their oaths and committed crimes against the laws of the land.

In parallel with efforts to restore vision and civility to the American nation, the rest of the world will also need to declare their own independence as part of a global effort towards declaring our mutual interdependence. What is at stake is a worldwide movement of transformation along the lines of what is proposed in the next chapter of this book.

Ecuador Today is an Example of a Growing, Functional Democracy. In 2004, Meredith and I moved from California to the Andes of Ecuador to establish Montesueños, a retreat center for peace and sustainability. What we hadn't anticipated when we first got here was that we actually were falling into a new cradle of liberty. Ecuador, like many other Latin American countries, had previously succumbed to the whims of oligarchs, military juntas, and puppets of the U.S. government and international bankers and oil, food and mining companies, but all that is changing now. We feel these positive changes in the air, a new sense of optimism and freedom from tyranny.

But in 2006, 80% of the people of Ecuador voted for Rafael Correa, who became its seventh president in a decade. Correa was different from the others. A Ph.D. in economics from the University of Illinois and with a background of volunteering to help the indigenous and poor, he formed a strong coalition of common people, the middle class and a large activist indigenous population.

Some American journalists and pundits believe it's unsafe to move to Ecuador because of this "leftist" or "socialist" platform, which now more represents the will of the people than the oligarchy and international business. Yet we've experienced just the opposite of these bogus warnings, which come from ignorant armchair pundits. For example, we're working with the Ecuadorian government to take strong positions on environmental issues. This struggle comes against great odds because of the enormous pressure to drill more oil and extract more metals, whose value of hundreds of billions of dollars keeps increasing in the long run. The temptations to do this are huge.

The new coalition is strong, spirited, and mutually supportive. The Ecuadorian people act not only at the ballot box in honest elections. They often show up in massive peaceful demonstrations to demand the resignations of presidents and legislators and policies that have clearly betrayed the public trust—actions now missing in the U.S. today. It seems many Americans have forgotten how a real democracy can function.

We feel a nascent democracy all around us in Ecuador. With more subsequent landslide votes of support, the Correa administration arranged for the public to vote for members of a constituent assembly to draft a Constitution for Ecuador throughout 2008 and a second vote to ratify the Constitution itself. What an opportunity for the will of the people to be heard! And the people said yes, it's official.

Many new policies are being considered such as placing off-limits public lands from oil drilling. The new government is also empowering communities to prevent oil and mining companies from continuing to destroy fragile ecosystems and habitats. The Constitution provides for an unprecedented new law stating that nature has rights. Hopefully, this will give us the needed clout to keep the oil in the soil, to stop the wanton mining, and to stop displacing poor farmers and indigenous peoples from their sustainable habitats. In September 2008, Alberto Acosta, who drafted much of the new constitution, came to Vilcabamba to celebrate the rights of nature. His speech was inspiring, and for the first time since I became an Eagle Scout 55 years ago, I felt chills of patriotism for my (new) country.

Correa has also declared the intention to close down the U.S. Manta Air Force Base and let the contract to expire in 2009. His administration proposed placing a large tax on oil windfall profits to go to the Ecuadorian people. They also intend to match funds with the international community to keep the oil in the ground in Yasuni National Park, one of the planet's most biodiverse Amazonian rainforests and carbon sinks, housing rare species and indigenous peoples that are voluntarily isolated. People in Ecuador are very excited about this country becoming newly unexploited and pristine under strict environmental laws, and becoming independent of any foreign control.

The peaceful democratic nature of political change in Ecuador should be an inspiration for how America could restore its own democracy through nonviolent activism.

Global Transformation and the Solution Revolution. The most important revolutions in history are not only about the struggles themselves to remove from power unsavory people and their outmoded thinking. These battles are just a transitory phase, hopefully to be followed by a new expansive worldview and system that can radically improve the human condition for a very long time.

For example, the Copernican Revolution 500 years ago positing that the Sun rather than the Earth was at the center of the solar system, formed the basis of Newtonian mechanics, which helped eradicate the ignorance of the Middle Ages and the exclusive Vatican stranglehold on knowledge and action in the world. Such new knowledge helped spawn the Renaissance, the Enlightenment, the Industrial Revolution, and eventually enabled us to travel into space.

The hard part of any revolution is during its beginnings, when the revolutionaries are persecuted for their contrary views. So it was for Galileo, who was imprisoned and Bruno, who was burned at the stake for embracing Copernican concepts. Yet if a revolution succeeds, a new paradigm can be born. [3] This struggle to change our basic concepts can be daunting.

We are now in one of these very challenging moments of history, seemingly poised at the edge of death and destruction on the one hand, and radical positive change on the other. I believe that if enough of us have open minds and hearts, we can pull this off. Necessity being the mother of invention, we can create a Solution Revolution of breakthroughs in having clean and abundant energy, water, food and forests. We can once again deliver quality health, education, research, infrastructure, and environmental protection in a peaceful, free society.

As a scientist who has stepped outside the box of Western materialism, I am aware of exciting new developments in having clean, cheap, decentralized energy which has been long suppressed by the powers-that-be, but which promise a powerful solution to global climate change and atmospheric pollution. [4]

But managing these kinds of new technologies will require the utmost care, as we place the responsibility into the hands of new social systems that we will have to trust. Therefore, it is a prerequisite that we create forums of discussion, positive intention experiments, and democratic discourse to decide which way we should turn to have a just, peaceful, and sustainable future, one that does not allow for the kinds of mistakes made in the past. As the advance of nuclear energy has so poignantly taught us, we cannot permit the people and policies now in power to continue to adapt new technologies to making weapons.

And so the stage is set for another revolution and unprecedented transformation. May the lessons learned about the debasing of America from its original structures and intentions provide us with the necessary paradigm shift in government away from war and toward the common good. May the ideals of the American Founding Fathers and Ecuadorian assembly of constituents provide inspiration for us to begin democracy anew.

The Solution Revolution can give us a new lease on life, free of entrenched interests and business-as-usual. With a collective newly-found respect for nature, we can free ourselves from the bondage of misplaced power and destructive practices, which are holding us back from implementing those solutions most of the world still sees as miracles.

Joseph Campbell said about our times, "Apocalypse does not point to a fiery Armageddon but to the fact that our ignorance and our complacence are coming to an end."

Ultimately, the Solution Revolution is a transformation in consciousness, of re-discovering the power of the void, of positive collective intention to create new worlds, not only motivated out of a sense of survival but out a sense of compassion, awe and wonder.

But our journey now mandates the difficult job of letting go of the familiar tyrannies of the past and present, so we can create the space to build bold and happy new futures.

References

(1) David Ray Griffin, *Debunking 9/11 Debunking,*
Olive Branch Press, www.interlinkbooks.com, 1996.

(2) Petition signed by prominent citizens on
www.dontattackiran.com.

(3) Thomas Kuhn, *The Structure of Scientific Revolutions,*
(Chicago: University of Chicago Press), 1970.
By paradigm, Kuhn means world-view.

(4) Brian O'Leary, *Re-Inheriting the Earth,* 2003, www.nohoax.com
and more recent essays on solution energy are posted on the
website www.brianoleary.com.

Portions of this appendix were first posted in October 2006.

Chapter 21
Taking the High Road through Consciousness and Combined Positive Intention

"You've felt it your entire life, that there's something wrong with the world. You don't know what it is, but it's there, like a splinter in your mind, driving you mad. You are a slave...like everyone else you were born into bondage. Into a prison that you cannot taste, or see, or touch. A prison for your mind."

– Morpheus, in *The Matrix*

"With knowledge comes responsibility...Neither governments, nor politicians, nor organizations, nor special leaders are going to create the changes that the world desperately needs. It is up to me, my neighbors, my co-workers—and all the other ordinary people all over the world—to learn how to work together and begin to build the cultures and societies that can enable us all to live fruitful lives in harmony with one another."

– Richard Moore, *Escaping the Matrix*
(the Cyberjournal Project, 2005-06, p. 188)

"The miracles of today will become the commonplace science of tomorrow. Maybe we can all become empowered to extract matter and energy from the void. These activities could bring us into a higher dimension which only awaits our acknowledgement and exploration...Then we will become more fully conscious of our place in a universe made alive and connected with who we are, and in the process, learn how special all life on Earth is. This is what the new paradigm is all about—transcending our self-imposed imprisonment by materialism, reductionism and determinism. We need to look at all this in light of the challenges of the twenty-first century...Perhaps we can now envision a sustainable future in which a blend of consciousness and common sense is creating powerful and benign new technologies whose time has come. We are going to have to support the science of consciousness…We don't only need a Los Alamos for new energy, we shall need to create a new Apollo program for consciousness science."

– Re-Inheriting the Earth (2003, pp. 220-221)

"The current discoveries inevitably lead to a revolution in the sciences which will spread quickly. Combining these new ideas along with some others now budding, is bringing us into a new paradigm, into refining our understanding of consciousness and the zero-point field. Perhaps this *is* the Consciousness Revolution…the miracle in the void is that we can all empower ourselves to create beautiful new worlds, magnificent new universes. When we begin to resonate with the majestic and ubiquitous reservoir of pre-energy and pre-matter in the zero-point field, we will all become healers, clairvoyants, and magicians. We can at last have peace, harmony, love and joy. Science is telling us that clearly, based on irrefutable experimental, theoretical and personal evidence. I invite you to trust the process and walk with me through the visible into the invisible."

– Miracle in the Void   (1996, p. 192)

"Intention appears to be something akin to a tuning fork, causing the tuning forks of other things in the universe to resonate at the same frequency…My own motive for writing *The Intention Experiment* was to make a statement about the extraordinary nature and power of consciousness. It may prove true that a single collective directed thought is all it takes to change the world."

– Lynne McTaggart, *The Intention Experiment,*
(Free Press, 2007, pp xxv, xxvii)

We finally come to the point of asking, how are we going to be able to thread the needle of free energy? If so few of us even know it is real, if so few of us know how to deal with this possibility as provocative supporters rather than selfish receivers who want to "get in on the ground floor" of power, money and influence, our society is obviously stuck in a matrix of our own making as prisoners of a not-so-subtle system of mind control. Even those seeking urgent sustainable solutions appear to be stuck within the boundaries of their own cultures.

Between 2002 and 2006, I taught a course in the Masters program in Transformational Psychology at the University of Philosophical Research in Los Angeles. Part of the intent of the course was to embrace all four cultures of the Phoenix. The title of the course was Science, Ecology, Ethics and Consciousness. The attendance was low, but the students that did enrol were among the most aware and sentient beings I have ever met. They began to understand how important all four cultures were for our future, and if we leave out any of these qualities and beliefs, or specialize too much in any one, we will box ourselves in.

For example, many Truth-Seekers have become so enmeshed in addressing the hidden agendas of the elite tyrants, they've been swept up in negative thinking. This negativity can become a self-fulfilling prophecy, giving too much energy to addressing what's wrong rather than what we can do. The intensity could give too much energy back upon the tyranny itself, which can be only a gift to the powers-that-be. These people can get immobilized by fear itself—fear of the implications of the dark agendas they have helped uncover, fear of being nailed as heretics.

Likewise, many Deep Ecologists also can be immobilized by their own fears of imminent planetary disaster. Again, we seem to have little hope if we listen exclusively to those of us that announce that the sky is falling. A part of me is sometimes too heavily invested in both truth-seeking and ecological warning signs, and I also realize it's tempting to simply wallow there.

The Pragmatists, on the other hand, are too tied up with their own agendas to be able to make a clean break towards embracing new paradigms. One notable exception is the free energy inventors themselves, whose knowledge and persistence require pragmatic qualities. Pragmatism is also the culture from which I have sprung, as a former mainstream scientist, so I am familiar with the dynamics, particularly when I begin to embrace the other cultures. For a pragmatist to escape the matrix is a daunting and lonely task...but these qualities are absolutely necessary to be able to move forward with free energy. This effort takes discipline and the willingness to be humiliated, ridiculed and debunked.

That leaves us with the Spiritualists, which in itself can be a loaded term. Pragmatic atheists and skeptics, for example, usually dismiss the Spiritualists as ungrounded airy-fairy New Agers or cult members indulging in fantasy. So I'm redefining this culture with the titles Consciousness Researchers and New Scientists. During the

1980s and 1990s, my research on new science revealed compelling experimental evidence for a consciousness that interconnects the entire universe across time and space. It is a force that is so powerful, it cannot be ignored any longer, in spite of propaganda to the contrary.

This mysterious energy field supersedes the four known (materialistic) forces of nature and can literally make matter, energy and information suddenly appear or disappear from our normal perception. It mandates higher dimensions of reality that can allow us to heal and transform everything.

My own evidence has been through: (1) personal experiences of remote viewing, healing and a near death experience; (2) outrageous demonstrations of materializations by psychics such as Sai Baba and Thomaz Green Morton; and (3) the clear results of experiments in quantum physics (the observer effect), experiments in psychokinesis and remote viewing, experiments of prayer and healing, experiments on altering and purifying the properties of water, and experiments with the zero-point energy field. These activities have become the mainstay of my own research over the past two decades, as reported in my previous four books, *Exploring Inner and Outer Space, The Second Coming of Science, Miracle in the Void,* and *Re-Inheriting the Earth.*

I define consciousness as our intention to create something new in the universe: to the degree the universe resonates with that intention and changes as a result, is the degree to which our consciousness is effective. Experiments show that combined positive human intention can alter the universe more significantly than we could do as individuals.

And this is where things can get tricky. It's also possible to use negative human intention to alter the material world. This is black magic, and we'd want no part in that. Fortunately, as reported in my earlier books, some experiments show that the forces of a negative consciousness appear to pale before the forces of positive consciousness. Some individuals attribute this phenomenon to a "higher power," and some others call it God.

This kind of discussion can become highly charged and grossly misunderstood. Nevertheless, some skeptical pragmatists look askance at *all* forms of "magic." There is also the inference that the "magical" effects of consciousness, even though it's been proven experimentally, should not be pursued. All this then gets confused with fundamentalism, evangelism, rapturism, and numerous other religions and cults, whereas, in fact, what we are talking about is an emerging experimental science.

The data clearly show us a recognition of a "spirit" which pervades all of nature, an awareness of the whole, a way in which we are all connected. The indigenous peoples have known this for a long time. This awareness can lead to a deep compassion for the commons. But consciousness research, new science and indigenous spiritualism are an anathema to those steeped in the zero-sum game.

I believe that the world needs to come together in a blend of inputs from the four cultures of the Phoenix, but only the "Spiritualists" or consciousness scientists can provide the lasting solutions. All other groups simply don't have the awareness to get there, but they do have an important role to play in presenting the depth of our problems. I wrote in *Re-Inheriting the Earth* (p.221):

"We are in a global spiritual crisis which demands that we remove our veils of denial and enter a new science of consciousness, exploring our potential to heal, our eternal nature, and our membership in a cosmic community of sentient beings."

The overwhelming evidence that we have been visited by technologically and culturally higher beings through the UFO/ET phenomenon adds another piece to the puzzle. It's my opinion that we earthlings need all the help we can get. If, through our consciousness, we appeal to the more sentient forces present in the universe, we might get the assistance we will need to escape the matrix.

That is my hope. We need all the assistance and educational support we can get to transform ourselves and our planet into an enlightened state of being—something most of us would be happy to embrace. Getting from here to there is now our task, and every great journey begins in small steps.

Meanwhile, here are some practical suggestions about how you might get involved:

1. Undertake your own research into new energy concepts, using the web, local library and taking relevant courses. I will be teaching many courses here at Montesueños.

2. Seek out likeminded individuals for a discussion group, both in person and on the Internet.

3. Hold public seminars and forums.

4. Become activists, develop alternative media contacts and learn about persuasion strategies.

5. Petition those members of governments who will listen. Actively work with governments that resist supprisive oligarchies and empires and that support clean energy, the rights of people and the rights of nature. Protest those governments that resist (e.g., impeach, elect new blood, strike, boycott, demonstrate, march, bang pots, obstruct aggression, form parallel governments, impose local rule and autonomy, etc.).

6. Read more and more. The Internet is full of important information.

7. Try out various experiments in consciousness and healing. Do spiritual practices such as yoga. Learn about the new sciences through experience, demonstrations and experiments.

8. Balance your fact-finding about the truth, the state of the world, with a basically positive attitude that you *can* make a difference. Remove yourself from fear-inducing environments and create your own peaceful and unpolluted enclave that encourages clear and ethical thinking, yet be willing to face malevolence when necessary (see the Epilogue).

9. Join the team! I'd like to hear from you!

I believe that the last four items are the most important. Because when you experience the energy solution revolution, you and the world will never be the same again.

Afterword
An Impossible World?

Arriving home from work, Appleby's heart skipped a beat when he saw the return address on the mail his wife had left on the kitchen table —perhaps this time he'd be offered a sale for his curious brand of fiction.

Excitedly, he tore it open:

"Dear Mr. Appleby,

Thank you for your recent submission to *Altair Space Monkey*. Unfortunately, we will have to pass on your story.

For your further development, comments from reviewers are included below:

'The entire premise of the story is that people could have an underlying predisposition to the pathological; i.e., a natural inclination to both superficialities and what might in more educated circles be politely referred to as a grotesque desire to dominate others. This is more than just a literary sleight of hand—it is the crutch of absurdity. I would suggest that if the author's intent is to merely shock, then the juvenile dime market might be more receptive of such efforts.'

And

'In times past, before the sciences of evolution, sociology and psychology became as advanced as they are today, your story might have had a chance—a slim one to be sure, but it might have been offered a 'revise and resubmit'. Nowadays, though, the readership of quality publications such as Altair Space Monkey has matured alongside scientific progress. Extrapolating from the known to the possible is fine, and indeed is the essence of good science fiction. But stories that leap from the possible to the most highly improbable do not engage our informed readers. Altair Space Monkey is a magazine of science fiction, not fantastic horror. The society described in your story would never progress beyond the Stone Age. Its people would rather spend their time gazing at their own images in rock pools than working together for any collective benefit. And if by some lucky spin of the cosmic roulette wheel they did produce more technology than just fire, such progress in science would surely be used in such adverse ways that the society you invented could never be achieved, let alone sustained. I would suggest that the author take the time to read an introductory text on evolutionary science.'

And finally...

'In your poorly conceptualised story, you make reference to a so-called 'modern,' 'technological' society that features, among other silliness, more weapons, more wars of aggression, more pollution, a steadily increasing number of working hours as well as increasing unemployment. Exactly how would that come about? You further suggest that the citizens of this fantasy prefer to spend their time chasing rainbows in the form of some sort of never-ending cycle of conspicuous consumption. You even have the protagonist utter that he must continue to do so 'just to keep up'...but who is it that these wretched souls are trying to keep up with? If everyone is buying the same things, then exactly how are they differentiated by what they own—anyone can just go out and buy more of the same, can't they?

'You also presuppose that no one in this society has access to universal power. That is, you seem to posit that none of the zero-point technologies we enjoy ever appeared anywhere—not even in a single garage workshop despite their simplicity—while billions in research dollars are being spent on ever-bigger Tokomaks, cyclotrons and other preposterously grand curiosities! Why would resources be ploughed first into scientific realms that, although arguably worthy of abstract attention, are unlikely to benefit the public on any tangible level? Who would allow such an illogical apportioning of resources? Or are you suggesting that they did develop practical alternative energy sources, but due to some strange hidden agenda, they were suppressed? Who would benefit from that? Again, such a society would very soon reach a 'boiling point,' a combined socio-economic-environmental crisis from which it might never recover.

'Think about it: if this insane world of yours, with over six billion people I might add, all self-chained to the same consumerist treadmill (or aspiring to be!) had to depend upon fossilized hydrocarbons for the majority of their fuel needs, the biosphere they depend upon would be overwhelmed in a matter of decades. And please, in any future submissions, do not attempt to paint nuclear power as even a borderline possibility. The majority of our readership is not scientifically illiterate, and hence is unlikely to present a blind eye to that little problem of half-lives of nuclear waste products (the safe disposal of which you neglected to describe). Given the environmental and economic ludicrousness of such an obvious hazard, why would any ruling powers—secret or otherwise—allow it? If this society were to collapse, as the sort you madly envision must, then what would such clandestine groups have left to rule over?

'Finally, in such a psychologically and economically barren world, what space would there be for art, literature? For a family? The average person—what you call 'consumers'—would be ground between the stone wheels of work and debt...'"

Appleby scrunched the rejection letter, tossed it in the trashcan. His face twisted.

"What did those twits know?" he asked himself aloud. Deep down, though, he knew they were right—it was the fourth rejection slip he'd received for this story, and the feedback was always the same.

But his imagination refused to relax its iron grip: what if history had taken a sinister turn instead of a sensible one? What if darker forces drove peoples' nature? He saw it still—an early world of resource scarcity, where some people—perhaps just one or two here and there—decided that it was better to look out for themselves first and worry about others later, and even then only if it was convenient, with a personal payoff. How might such individuals, who would surely grow in number given the starting conditions, evolve socially and continue to prosper?

Appleby's mind soared across the millennia, to a society where self-interest and struggle for control of others was the norm, cooperation the deviation, where citizens competed for more baubles, more new shiny things to declare their status over their lesser brethren. Rats in cages, spending their pitiful lives dashing about on exercise wheels going nowhere, the sputtering candles burning low, while from the shadows darker powers watched in amusement.

It had to be possible.

Emboldened by his imagination, Appleby dashed off another submission letter.

Afterwards, as he sat sipping a coffee, his gaze drifted to the toaster-sized ZP inductor in the kitchen, and he thanked his lucky stars that he lived in a world where he never had to pay a power bill, where heating and cooling were free of energy costs, where his water was pure and veggies local and fresh, and where a four hour work day allowed him ample spare time to indulge his writing fantasies.

Outside, the air was fresh and the streets clean. Everyone had a job, and in this world—the real world—there were no clandestine powers holding back progress. All received a fair share for their efforts.

For a moment, Appleby wondered what the citizens in the gloomy world imagined in his story might make of this one? Would they too consider it impossible? A utopian dream beyond their grasp?

He reached for his pen.

****

To best understand Dr. O'Leary's latest book *The Energy Solution Revolution,* I invite you to join me on a short detour through the fascinating world of game theory—it won't take long, and I promise it will be worth it.

From the perspective of "game theory," human history has been characterized by a continuing struggle between self and collective interest. The first comes naturally, for as the psychologist Donald Campbell once said, it's not necessary to teach people to look after themselves; they work that out pretty quickly. In game theory, looking out for one's own interests exclusively is labelled as a "zero-sum" game, where the outcome is framed as a competitive win-lose scenario. That is, for me to get ahead, you must lose.

Obviously, early man would never have got very far socially or technically employing such a one-eyed approach to life. Indeed, Campbell reported a conflict in all traditional cultures between those values and attendant cooperative decision-making processes that are supportive of the group or tribe—also known as "non-zero-sum" or win-win games— and those that benefit the individual specifically. Hence, it was understood that although both were necessary there was an urgent need to keep them in balance.

According to science writer Robert Wright, in aboriginal cultures, non-zero-sum games acted as an insurance policy against lean times. When an individual gave excess food to those who needed it, the "IOU" engendered by the act provided a buffer against possible future hardship the giver might face. That is, in the absence of a fridge, food was best stored in the guts of others, and moreover, being in a position to grant such boons immediately conferred status to the giver. And seeking social status is at the crux of game theory: once food and basic survival are no longer an issue, something else must take their place so that individuals can assert their distinctiveness and superiority:

namely, prestige symbols and conspicuous possessions. Whether we're talking about glittering glass beads or designer "investment" handbags, the underlying objective is to broadcast loud and clear the same simple message: "Here is my position in the pecking order—I am important."

The need to manage surplus resources and exchange IOUs for non-essential prestige goods brought with it social complexity. For starters, there was tension over how to divvy up the costs and profits of collective enterprises. There was also the need to manage the exchange of IOUs, and keep good relations wherever possible with other tribes. And when inter-tribe relations broke down—often when one side got a little too greedy with the exchange of those IOUs—someone had to rally sufficient numbers to fight for the survival of the collective. Enter the leader, the village "Big Man," someone with the people skills to rub the necessary oil to smooth inevitable intra and inter group frictions, and the economic acumen to ensure that his people benefited more from cooperation than they would from going it alone.

Obviously, anyone who fits that job description deserves to be rewarded, whether it is extra wives or a fat annual bonus and company car. The more cooperation they can encourage, the more excess there is to be traded, more prestige symbols to be shared. So, even though the leader, as befits his position at the top of the pecking order, has first pick of the trophies, everyone wins: the system works *only* so long as the net collective gains exceed possible individual gains. It's not surprising then, that animals such as humans are hardwired by natural selection to despise cheats; those people who break the rules, ignore IOUs, or take an unfair cut from exchanges.

Now, it stands to reason, doesn't it, that such a leader—that "Big Man"—would be both highly visible and very accountable for the welfare of his collective. As Campbell noted, the Big Man's job is to find a balance between self and collective interests. And as far as most people are aware, that's probably the way it's always worked, from the first hunter-gatherers, to the modern era of Wall Street Traders. Everyone gets their share, and the Big Man—and Big Woman—will carry out their job to make sure that system of equitable distribution of cooperative surpluses continues. In the parlance of game theory, where each non-zero-sum collective game hides a zero-sum individual game, those at the top get the most, with returns diminishing as we move down the ladder of status, but each individual necessarily must gain more than if she were to go it alone.

Indeed, for many people, as Dr. O'Leary points out, it is far more comfortable to accept this idealistic belief than to question it. After all, what could possibly be wrong with our modern consumer society, the pinnacle of human accomplishment?

Well, for starters, in stark contrast to the wisdom of our forebears, in modern times individuals are told to look after themselves, constantly and immediately. Look around you: on the television, in print media, even on the handy-teller screen, someone is telling you that YOU matter, and that YOU deserve IT, with IT depending on what baubles the spruiker is selling. Self-interest is the rule of the day, and the spruikers make it all seem extra special. It's as though everyone else is eating it, or wearing it, driving it, or wiping with it, and if you aren't then there's something a little odd about you, *less* about you, because you're not keeping up with the person next to you. After watching these ads, many have the feeling that they're missing out on what should rightfully be theirs, and because of that they're not quite as good as everyone else. Until they buy it, and then until it gets replaced by the next, newer, glitzier IT.

So, strike one against the Big Man for urging us to pursue individual wants at the expense of group needs. In fact, it is not too much of a stretch to say that in modern consumer culture, groups of "one" are fast becoming the norm, with individual decision making that is often skewed to, well, the individual, instead of collective concerns such as the environment and the biosphere that supports us. Compared with the long, gradual arc of human history, where cultures have out of necessity become more complex, "going it alone" can only be considered counter-evolutionary.

What else did we say is expected from the Big Man? Honesty and visibility. Although he gets the biggest cut, everyone must still gain more from the collective interaction, not less. After all, we're hardwired to hate cheats. So, we expect that any technological innovations that become available—whether they be fish traps or cleaner sources of power—will be promoted and distributed in such a way that benefits everyone, not just the Big Man.

Unfortunately, according to Dr. O'Leary, this is the second strike against the Big Man, those leaders in whom we trust to do the best by us. It would seem that there already exist workable solutions to many of the problems currently facing this planet in the form of sustainable, non-polluting energy technologies that could power those fridges our ancestors didn't have and the plasma screens we now crave. While our individual interests have been turned away from the collective prize, our eyes glazing over at the prospect of the latest shiny "new" baubles, something very disturbing has happened. According to Dr. O'Leary, the "Big Man" of old is no longer visible, and no longer accountable. While our eyes were off the game, new players have entered the field, and their strategy of withholding much-needed technological advances is decidedly zero-sum.

In the middle of the night, while we slumbered, blissfully unaware, the new Big Man has stolen the prize away from a mesmerized commons. In its place is an Orwellian nightmare of control and impoverishment of the masses. Nowhere is this imbalance more evident in our (that is, the Big Man's) exploitation of Earth's resources and the insane choice of energy sources. Double-speak is the common slippery tongue. Coal has become "clean," nuclear "green." Those who dare to question the new order of things are quickly labelled "conspiracy loonies," no matter what their background or what they have to say. Such dangerous thoughts, we are told, must be given no thought at all...

In such a topsy-turvy, upside down modern world, what becomes of those who seek to improve the lot of society, those with the imagination and the intellect to think of brave new non-zero-sum worlds? Perhaps my short stories can provide some insight to that question, and to Dr. O'Leary's central premise in this courageous and much needed book.

Regardless, civilizations decline when non-zero-sum gains become increasingly centralized into the hands—and pockets—of the powerful. Hence, the basic foundation rule of a non-zero-sum cooperative game—that all players receive more individually for their collective efforts—is broken.

As result,

1. There is a weakened incentive to work for the good of the collective (and remember also that in modern consumer society, individuals are already told—over and over again each and every day—to look out for themselves).

2. There is also increased social unrest, as those individuals begin to realise that 'all is not quite right,' and that they're in fact not getting their fair share...

3. Part of this stems from the heavy tax burden needed to support an increasingly top-heavy and expensive bureaucratic system, which in itself is required to:

4. Surveil and monitor the mass of not-so-happy individuals that comprise modern society. The functions of surveillance and monitoring do not in and of themselves contribute positively to either social or technological progress in civilization—they act as damping mechanisms and restraints...

All of the above points combine to reduce individuals' incentives to add to the collective in a meaningful way in the form of new ideas/technologies/simple manual effort.

Therefore, the system gradually collapses (and yes, while the tax collectors are still 'fiddling).

– Shaun A. Saunders

Epilogue
Open Appeal to Create a World of Sustainable Abundance

How to face Resistance to an Energy Solution
Revolution and a Plea to President Obama

"(Geoengineering the climate) has got to be looked at. We don't have the luxury of taking any approach off the table." (author's note: how about developing free energy instead?)

> – Dr. John Holdren, President's Science Advisor,
> Associated Press, April 8, 2009

"Today we the people are starting to understand our banking and monetary system, and we are shocked, dismayed and furious at what we are discovering. The wizard behind the curtain turns out to be a small group of men pulling levers and dials, creating an illusory money scheme that, behind all the talk and bravado, is mere smoke and mirrors...it's time for the government to print our money, like the Greenbacks of Lincoln's time."

> – Dr. Ellen Brown, www.globalresearch.ca, April 9, 2009

"Any government that can disburse $2 trillion secretly, without any accountability, is not a democratic government. It is government of, by and for the bankers."

> – Chris Powell, "Fed Refuses to Disclose Recipients of
> $2 Trillion," GAIA, Dec. 12, 2008

"The Great Ponzi scheme that is the Western World's economy has grown so big there is simply no fixing it. Flushing more debt through the system would be like giving Madoff a few billion to tide him over. Or like adding another floor to the Tower of Babel. To what end? The collapse is already here. The question is: how much do we want it to hurt? Using the public's purse to finance "confidence" in a system that is already kaput may delay the Day of Reckoning, sure, but at the cost of multiplying our losses. Perhaps fantastically."

– Rolfe Winkler,
"More Debt Won't Rescue the Great American Ponzi,"
Option Armageddon, March 9, 2009

"Let me cut to the chase. The biggest robbery in the history of this country is taking place as you read this…Bush and his cronies—who must soon vacate the White House—are looting the U.S. Treasury of every dollar they can grab. They are swiping as much of the silverware as they can on their way out the door…From talking to people I know in D.C., they say the reason so many Dems are behind this is because Wall Street this weekend put a gun to their heads and said either turn over the $700 billion or the first thing we'll start blowing up are the pension funds and 401(k)s of your middle class constituents."

– Michael Moore, www.michaelmoore.com, Sept. 29, 2008

"Both Western Finance and Western Medicine are fundamentally based on fraud…The end of these systems is now in sight. They are crumbling under their own arrogance and stupidity, revealing a society based on self-righteous deception and global scandal. Everything we thought was real turns out to be fabricated: The money, the medicine, the economy, the law…it's all being revealed for what it is: A *Matrix* of enslavement, designed to keep the People believing they live in a free society, even as their health and wealth are stolen from them by the sinister few who wield political power…Wealth is not a collection of digits in a computer. It isn't a promise printed on green paper money. Real wealth is a garden that feeds you, a river that hydrates you, and a system of medicine that nourishes and supports you. Real wealth is a day with sunshine, a night under the stars and a life lived with purpose."

– Mike Adams, http://www.naturalnews.com:80/024353.html,
Sept. 29, 2008

"Every so often civilization seems to work itself into a corner from which further progress is virtually impossible along the lines then apparent; yet if new ideas are to have a chance the old systems must be so severely shaken that they lose their dominance."

– Chester G. Starr, *A History of the Ancient World.*
Oxford University Press, 1991 (p. 124)

In Shaun Saunders' ironic twist on reality, the authentic world he envisions is a world of abundance whereas Appleby's fictional world of scarcity and control is rejected as "impossible."

But the facts of today's real world *are* the fictional world mocked up by Appleby, and the real world surrounding Appleby cannot even be imagined by most people.

This is *crazy* but true: The world we now inhabit *is* a world clearly headed towards totalitarian destruction, yet the world we could have—one of great abundance--waits in the wings for when the ruling big men and the obedient masses let go. Then we won't need any big men except for ones benevolently piloting Spaceship Earth with compassion. It is difficult to grasp this potential reversal of our fates at a time of mass hallucinations with which enough of us can trust tyrants to guide us into such distraction and despondency. Yet that is what is happening.

So what's going to happen to us and when? I predict we'll know a lot more about the collapse of Western civilization before you read this. The ship of state is sinking so fast, any speculation about the next disasters will undoubtedly be inaccurate, because We the People don't yet hold the cards. So why bother to guess?

What we *can* know now are the dynamics underlying the collapse and the need to get back to basics regarding our social and individual health.

Saunders echoes Mike Adams of *Natural News* that our financial and health care systems are two falling pillars dominated by money and greed. He also feels that the third pillar should be the timely introduction of free energy—which could have happened more easily sixty years ago, when things were more optimistic and when the wrong big men didn't have things so obviously sewed up and messed up as during these times. I agree. Saunders and increasing numbers of us can pretty easily guess in broad terms what is really happening behind the scenes, why the rulers have kept us away from free energy for so long. He wrote to me about three possibilities:

(1) Some in control do know it is needed but simply do not know how to introduce it without severely affecting the existing financial/power paradigm (i.e., it's in the 'too hard' box);

(2) the 'warhawks' don't give a stuff anyway about the environment, and simply wish to continue squeezing the populace for more and more power (eventually resulting in a world that will make *Mallcity 14* look like a summer camp); and

(3) the big men are privy to a possible forthcoming world-wide catastrophe, and in the meantime, it is 'easier' to allow business as usual, and afterwards, when the chosen survivors come out of their black-budget bunkers, they will start anew with all the technology available.

Saunders and Frazier also both speculate that an ET intervention, whether real or staged, will bring the stakes ever higher as the drama spirals ever upward and the elites grab their lifeboats while the *Titanic* goes stern-up (e.g., http://www.ahealedplanet.net/ufo.htm).

But the most vexing question of all questions is: how can otherwise intelligent, enlightened and compassionate people *allow* such malevolence to sink us all? How can the scientists, environmentalists and progressives continue to be oblivious to the fact that we have already hit the iceberg and we need to do something very quickly to man the available lifeboats and miracles with consideration for everyone? Wade Frazier wrote to me:

> This is a conundrum like no other. Nobody has cracked that nut, yet. As I discuss, the "faith" of "liberals" and radicals does not allow for the possibility of conscious manipulation of the system by those in control. There are at least a few reasons for this:

1. *They do not understand spirituality.* They project their understanding onto others, thinking that everybody shares their perspective and motivation. There is a spiritual dark path, and people walk it, and many of them are in positions of power. And my point for those doing the projecting is to understand that there are "evil" people in the world, and that they do not think like the rest of us. Few of us (less than 5%) are on the spiritual dark path, but for those that are, they approach life differently than the rest of us do. When we tell lies, our conscience, to one degree or another, kicks in and makes us uneasy, and we usually try to do better next time. Being in denial of the lies is one of our defense mechanisms, but at one level or another, we are aware of the lies, and it does not make us feel good (the main reason for the denial). For a dark path person, telling lies and getting away with it is a triumph. The more people that are duped by them, the larger the victory. They have a different scorecard than the rest of us. Not understanding that they are different allows people to deny that they exist.

2. *They are wedded to the structuralism of the Chomskies and scientists of the world.* Science is obsessed with finding the mechanism, and completely ignores that maybe something or someone designed those mechanisms that they are so keen to discover and describe. They study Creation, and deny that there may be a Creator. I am not into matters of faith, but materialism is a faith, too. Science is

a religion for most scientists, and it has its same heretic-punishment, defenses of the faith and other aspects of organized religion.

3. *Very few have sufficient personal integrity to investigate and accept the issue, and experience is the greatest teacher, and until they actually have some experience of how the system really works, it all sounds like another theory to them.* To actually understand how our world really works, and how we are being screwed by the very people, institutions and "faiths" that we gave our power away is to come to a place of responsibility. 99.9% of people would rather keep playing the victim than accept responsibility. Those who obsess over the "conspiratorial" behaviors are not accepting responsibility, either. As I emphasize over and over, the path to our salvation is not taking on the dark path folks who we gave our power away to. It is taking back our power, and doing it gently. If people ever overcome their denials that those they have given their power away to are screwing them with it, they then want to go "get" the "bad guys." That is no answer. Love is the answer, and always has been. Now, I will allow that it is very possible that we are here to play this game of giving our power away and being screwed over by those we gave it to. Many bodies of mystical material allude to it. It may be that our souls want to play this game of kill and be killed, and "I have the power and you don't." Maybe that is what this dimension is all about. Now, if that is true, I really question my soul's wisdom, and it is easy to get quite angry with whoever set up such a game. It may be that we are about to finally learn the lessons that we came here for, and just in the nick of time, because we are about to destroy humanity and the planet, through our many failings, with the many ways that we allow fear to manifest being chief among them.

So, that is a tough conundrum. One of the most amazing things is truly understanding how deeply the fear and denial are rooted in 99.9% of the population. They *do not want* to understand. Their hearts are not open, so consequently, their minds are not. You can lead a horse to water, and all that. Part of the problem is also having tunnel vision. About 95% of the population is scientifically illiterate, and has no idea of how the world really works—they don't understand those mechanisms that scientists are always pursuing. People can put their feet on the gas pedal and pump gas at the gas station, but they have only the faintest glimmerings about how it all works and that they are filling their SUVs' tanks with a year's worth of calories for their bodies. For those who want to understand, it is possible to do so, but they have to *want* to understand, which very few really want to, because the implications are either overwhelming or their souls want to keep playing the victim and learning the lessons of fear, so they flee from the implications. The more sophisticated of us act like they are giving work like mine a fair hearing, but it is just a game that their egos play.

Until people have their own personal encounters with (dark-path) people, in a way that their motivation becomes crystal clear, they will tend to not believe the motivation behind what is happening... I think that people have to have personal experience with those kinds of people taking their masks off, to really "get it." And the dark path people know this well, which is why somebody like me had to get hammered repeatedly before really getting it. This is one of the key parts of the conundrum, and one of the hardest to understand. For the other 95% of us, we would rather not think that people can be that way.

In other words, it seems most all of us need to have the *experience* of facing the darkness before we can see the dawn. There appear to be no easy pathways to achieving this kind of sentience. The darkness is really there, and to deny it and not assign responsibility to those who are suppressing a bright future is not taking our own responsibility to re-choreograph our future. The challenge becomes more serious: it's tempting to say, "if I didn't directly witness or experience free energy or its suppression, it can't be true."

But such denial can actually be dangerous. In Martin Luther King's words, "To ignore evil is to become an accomplice to it." *Embracing the possibility of a breakthrough energy solution revolution should be our mandate, not an idle speculation to be ignored or ridiculed or feared.* To thread the needle of free energy, a critical mass of us, combined in positive intention, will need to muster the courage with open hearts and take this firewalk—together. This may well be the central mystical theme of the crossroads the world now faces. Again, Wade Frazier:

I believe that free energy can only be pursued by the fully sentient, or those closely so. I think that is the intent of whoever set up this earth game. As you know, people at a high level of sentience are extremely hard to find these days, which is part of the conundrum. So, my approach has been to seek people who genuinely seek the truth and solutions, and give them something to chew on. I had not seen that approach tried yet, which is why I ended up doing my site www.ahealedplanet.net.

The Lone Rangers of free energy get picked off one at a time, like ducks in a shooting gallery. If they can overcome their own limitations to the degree where they try to mount any kind of effort, their allies usually present more of a hazard than the Big Boys (aka, the Big Man) do. There is currently not enough collective integrity in the masses to overcome their inertia and the organized suppression, and almost every activist group I have ever heard of hacks at the branches of the issues and is hooked on their particular scarcity-based way of viewing the world.

So, an untried avenue, at least as far as I saw, was just trying to help the awake and awakening see the big picture and where the primary leverage point is: energy. If they can just understand that and how the world really works, we may be onto something. Although time is very short, I think that any effort that attempts to go straight from ignorance to storming the free energy Bastille (or with a brief interlude where we collectively nod and delude ourselves that we have the right stuff) is doomed from within and easily defeated by the Big Boys. It is not easy to grok the free energy milieu and conundrum. I am only seeking to help people see that picture.

If a sizeable group (probably several thousand) can get that far, and truly let go of its scarcity-based beliefs, at least while pondering the free energy milieu, then we might have a chance to get active from there. Again, I have seen almost all the best and biggest names of American free energy activism wave the flag, and I am doing my best to get beyond all scarcity-based thinking, or at least point it out when it rears its head. I think that, because all earthly groups currently promote their favored brand of scarcity-based consciousness, people try to pander to it to get their foot in the door. I think that strategy is doomed from the outset. For that reason, approaching any group may not be the way to go, but those thousands will come from many walks of life. Heck, nothing has come close to working yet, but this at least seems like it may have a chance, although it is truly looking for needles in haystacks.

The dilemma here is that, there may not be time enough for a sufficient number of us to go through this kind of sentience-and-free-energy-awareness "training program." To those of you who haven't yet taken these steps, I implore you to suspend disbelief and to embrace the *possibility* that an energy solution revolution could save us all, and that you could really help us. What would you have to lose in giving this a try? Think about it: In light of what you now know, are fear, apathy, pride, credibility, and fictitious self-interest over the common interest, still more important to you than looking for all reasonable pathways toward averting a planetary disaster?

With the recent election of Barack Obama as the U.S. President and the deteriorating economy, the rules of the game are changing. Energetically, the Obama victory came just in time to relieve us from a collective nervous breakdown. In that way, it's most welcome to get rid of the most genocidal/ecocidal criminals of all time, thugs who have been totally in charge of cutting us off from our most heartfelt visions.

But my tears of celebration are bittersweet. Our journey has really just begun after eight-plus years of a U.S.-led global cardiac arrest.

How can we possibly get our societal policies to align with our potential for an energy solution revolution if even the most progressive among us cannot even fathom its possibility? How can we get beyond the gatekeepings of Obama's inner circle of such vested establishment advisors as Chief of Staff Rahm Emanuel alongside the money lavished upon him by the Wall Street moguls and military-industrial global chauvinists who have no interest in something that threatens their power?

Put differently, will the Obama administration be able to clean up all the corruption, develop the imagination and vision to be able to think outside the box, *and* have time left to peel the onion of free energy in today's climate of limited thinking and competing priorities? Barring a miracle, I highly doubt it, given the disappointments I and others have so far encountered with the most publicly visible alternative energy progressives such as Ralph Nader, Dennis Kucinich and Al Gore. But it's always worth a try, and the massive energetic support the people of the world feel for "change" could open Obama and the rest of us up to unexpected paradigm shifts.

Besides the hopes of an Obama presidency, we have an additional argument to develop breakthrough energy as soon as possible. In today's climate of fiscal austerity and rampant indebtedness, it is unlikely anyone would be able to come up with enough capital to fund a pervasive solar or wind energy economy, least of all, the bankrupt governments of the world. The trillions of dollars needed are simply not there. To meet today's energy demand, we'd need on the order of $20 trillion just to install materials-intensive wind and solar systems to replace current sources.

Therefore, only free energy could satisfy both environmental and economic criteria for a sustainable and abundant future. An awakening of awareness, first among some honest and sentient scientists, environmentalists, progressives, and many of the rest of us at the grassroots level, will become absolutely necessary for us to have an energy solution revolution.

The April 19, 2009 CBS "60 Minutes" mostly-positive documentary "Cold Fusion is Hot Again" is an encouraging sign of a reversal of fate in our dismal cultural perception of solution energy. This media-spun "redemption" of cold fusion is itself presented as the main story rather than the consistently-known but denied reality of breakthrough energy prospects, which the mob of ignorant mainstream scientists and reporters have debunked as fraud for so long. It's amazing that for twenty years now, Drs. Fleischmann and Pons have become so humiliated in exile while their pioneering results still sit in a controversial limbo of paltry support. They deserve the Nobel Prize, not ridicule, and Mr. Obama and the rest of us need to get on the case—pronto! Nature cannot wait any longer; how can we imagine neglecting *real* answers that cry out for responsible deployment *now?*

My friends, it's up to each of us to apply the pressure where and when it's needed most. Always. The election of Barack Obama has re-energized activists to do just that, relentlessly. May the open appeal to Mr. Obama below—which is really an appeal to all of us —be a beginning

And so, dear reader, we conclude this book with a question for you. Whatever awareness you may now have about these things, would you be willing to embrace a future world of free energy and other sustainable breakthroughs? Could you make it a personal matter for you to do so? If so, let's join up.

– Brian O'Leary
Vilcabamba, Ecuador
April 2009

Universal appeal to Mr. Obama and all concerned people from a fellow achiever who is gravely concerned about our future

Dear President Obama and the world community,

Mr. President, I congratulate you on your election. As a former astronaut, Eagle Scout, Ivy League professor, frontier scientist, futurist, advisor to presidential candidates, and an international author and speaker, I can identify with your feeling of significant motivation and achievement. But, in my later years as I now approach the age of seventy, many of those perks of recognition pale before a sense of urgency with which I feel we must approach our task of transforming humanity and nature into an environmentally, socially and morally sustainable future.

I honestly don't know whether we'll make it through these times. You're in the driver's seat now, and your task is daunting. You will have to stand up to some very powerful interests who do not want to live under those public policies that will become necessary for us to survive these times. You must understand this most basic conflict of interest in your position. But are you aware enough of what is really happening to us all? Do you truly know the depth of the crisis and the breadth of opportunities that lay outside the box of conventional thinking? I must admit that during my days of relative fame, I was largely oblivious of the deeper issues before us. The spotlight itself has a way of distorting our perceptions of reality.

In today's world, there is so much suffering, ignorance, neglect and corruption. Leaders nowadays prey on this condition. By supporting some of the most criminal actions in human history, the powerful elite have created an atmosphere of mass obedience by a fearful and helpless populace to wanton genocide and ecocide. We are destroying ourselves and each other and nature through the selfishness and greed of the few. As a result, unrest is brewing in response to monumental military, economic and political tyranny.

You must know that true knowledge, wisdom and compassion are threats to the status quo. George Orwell said, "In a time of universal deceit, telling the truth becomes a revolutionary act" and Isaac Asimov wrote, "When stupidity is considered patriotism, it is unsafe to be intelligent." I believe we now live in such a time to the extreme, and this situation is particularly poignant for us Americans now living under the tyrannies of empire and economic collapse. You come into office on the wings of conflict between a private power that has incubated you and a public clamoring for authentic change which so many of us have entrusted in you. You stand in the middle of an enormous gulf of interests, and you must know that at its deepest levels.

Those of us who take the road less traveled towards a greater truth always have been, and still are, martyrs placed on the altar of change. These heroes, often unrecognized in their own time, become the objects of religions and nation-states, which then often become self-aggrandizing dogmatic institutions that have nothing to do with the original intentions and spirit of the founders. The founders could only be rolling in their graves about how their contributions have become so distorted.

Continuity and bipartisanship cannot be nearly as important as returning to the basic principles of integrity, civility and public trust. You'll need to confront the tyranny of our recent past. Sacred myths such as the official stories of the JFK assassination and 9/11, for example, become enshrined in a fog of deception, lies and ignorance that distract us with bread and circuses, which, if perpetuated, can only precede the inevitable fall.

You must know the truth of all this, it's too obvious for you or any other intelligent and sentient being not to be able to recognize. You must know that we have an imperiled planet and civilization that cannot endure the collective neglect of humanity. You must know that we cannot rely on half-measures such as a slow withdrawal from Iraq, miniscule reductions in the defense budget, insignificant nuclear arms reductions, bank and Wall Street bailouts, or advocating nuclear power, "clean coal" and carbon cap-and-trading as lasting remedies to climate change. You must know that such minutiae cannot solve these problems. You must know, at some level, about Einstein's dictum that no problem can be solved at the level at which it was created.

Our species has invaded our home planet with such cancerous

vengeance and with such little conscious awareness or acknowledgement of the depth of our dilemma, it is hard to imagine how we can
get out of this matrix. But get out we must. It is much too late for us
to fulfill your mandate to "change" in the way you have embarked on
choosing your advisors and cabinet members. You must know that
these individuals are throwbacks to prior administrations who are the
epitome of an old ideological paradigm of a rule that cannot work in
these times. Many of us have become suspicious that you are only
paying back those elite individuals and groups that paved and paid
your way towards where you sit now. Have they so threatened you to
conform-or-else that you can't act differently?

For you to succeed you're still going to have to bite the hand that
fed you. You will have to feint your way as in a basketball offensive
around and through the broken field of defending opponents, who
include those who have supported you so well financially. You will
have to stand up in your courage and "betray" them (from their point
of view).

Are you up to the task? Are you willing to risk life and limb to
lead us into taking those actions needed to create a viable system of
governance? Are you motivated enough to join the ranks of our
brothers Martin Luther King, Mahatma Gandhi and John F. Kennedy
to take bold actions toward a peaceful, just and sustainable future for
humankind? Are you willing to evict the money-changers and militarists from the temple?

At some level, I think that you're aware that your definition of
"change" in no way resembles the kind of "change" any sensible and
knowledgeable person not beholden to vested interests would feel is
truly necessary for our own survival.

Mr. President, we're all in great trouble if you are unwilling or
feel unable to address the pleas of the vast majority of the people of
our nation and of the world. They cry out for peace, sustainability and
justice. You must also address the fact that nature, too, has rights.
Reforming the voracious appetites of the moneyed interests and of the
military-industrial complex will not be enough. Look all around you
at the degraded environment, at the suffering of the peoples of the
world longing for food in their mouths and for the kind of leadership
that could relieve their pain and give us all a reasonable chance to
move forward.

The crisis of America is first and foremost a moral crisis that has
moved us away from the fact that we have a physical situation that
demands physical solutions to have even the *possibility* of peace, sus-

tainability and justice. We must now stop the wars, stop the torture, stop the criminal corruption and the lies, stop the theft of the commonwealth, stop the surveillance of the innocent, stop all the polluting, stop the suppressions of true innovation. You must courageously lead us into a state of truth and reconciliation of our culture and stop the acting and pretending that authentic change is what you say you intend to create, because that isn't what you're doing—yet anyway.

I suggest to you that we need to define what is meant by "change." The change you could believe in and you have often expressed is what we might call *incremental change.* These are the small rhetorical feel-good kinds of changes that separate Democrats from Republicans, the liberals from the conservatives, the Tweedledees from the Tweedledums, all inhabiting a narrow spectrum of ineffective "centrism," holding onto power for dear life as Rome burns. These are the kinds of changes that got you elected as you navigated through the narrow passage between the special interests and the appearance of a public interest.

Beyond all that, we could consider what one might call a progressive *structural change,* such as eliminating electoral fraud, serving justice upon past criminality in high places, reregulating Wall Street, restoring the Constitution and the rule of law, re-establishing the "real economy" based on productivity rather than gambling away and squandering the public treasury, controlling the excesses of the Pentagon and an imperial aggressive foreign policy, restoring to Congress the power to print money and declare and fund wars, and allocating more resources to health, the environment, education and infrastructure.

Many progressives are desperate to restore this kind of common sense at the structural level and to create another New Deal for the economic crisis. The liberals would be grateful if all that were to happen, just to get us out from the deep hole we now find ourselves in. They would be satisfied to go back to the Clinton and Roosevelt years, to have just a bit more common sense in a world-gone-mad. Even some degree of neoliberalism, or ecomomic globalization (i.e., exploitation and biocide by other means), might seem sort of OK, in this view. Yet we know these measures are not OK legally or morally; they are only the actions of economic imperialism.

But in today's world, you must know that even structural changes in and of themselves will produce too little too late, and may be counterproductive in the long run, as we again become lulled into a false sense of security and buy a little more time before the inevitable collapse. Whereas incremental changes address mild corrections that really don't amount to much, structural changes look at how the current system can be modified to bring things back to where they were in somewhat better times. These approaches can only give us a frame of reference to launch authentic change. What we must have is *systemic change* to an entirely new paradigm of governance in the public interest that truly addresses the challenges of our times at the level they will need to be met.

Structural change cannot truly answer grievous violations of the public trust nor can they answer today's deepest issues. Only systemic change can do that. Whereas structural change can relieve the stress of a crisis in the short term, it cannot survive the test of time. Structural change can restore some sort of sanity to our systems, it cannot address the systems themselves. We must now challenge the precepts upon which our political, economic and military systems are based. We must deeply question the "isms" upon which we depend such as capitalism, militarism, neoconservatism, neoliberalism, centrism, monetary socialism, economic globalism, Zionism, terrorism. A new era must dawn, a new set of systems will need to be put into place in the near future for us to survive. Can you in your heart agree with what I suggest here? Or will you deny the gravity of these problems from your high and removed perch? Will you solely rely on probably misinformed and outdated advice coming from your cabinet and your staff? For you to advocate and map out systemic change, you will need all the help you can get from other quarters.

What would a world of positive systemic change look like? Generally, it should have the following features in which you will need to take the lead:

1. *Restore the letter and spirit of the Constitution and Bill of Rights.* This is the most major structural change we should do immediately.

2. *Fearlessly initiate a program of truth and reconciliation,* overseen by a jury of citizens without vested interests in the current system. Truth-telling cannot any longer be dismissed as conspiracy theory. The greatest conspirators are now holding all the political and economic cards and they must be exposed, whether it be electoral fraud, excessive private campaign financing, illegal surveillance, torture, illegal war, false flag operations, pollution or the embezzlement of the public treasury. All these things, and

many more, will need to come to the light of public scrutiny. The process of reconciliation seeks to return us to the rule of law and to serve justice upon those who have violated it, with fairness and compassion for all.

3. *Dissolve or stop funding those influential institutions with agendas that are blocking change* toward global peace, justice and sustainability. Start over. Dismiss the leaders of most of our public institutions and build new ones from the ground up. Stop funding those private institutions that dip into the public treasury in ways that are clearly immoral and unproductive. This will require a courageous stand to dissolve the current federal bureaucracy as it is (Dod, CIA, NSA, the current treasury, the FBI, Department of Justice (sic), Department of Energy, etc.). Expose international institutions such as the Illuminati, the Council on Foreign Relations, the Bilderberg Group, the Trilateral Commission, World Bank, International Monetary Fund, World Trade Organization, the Federal Reserve and other central banks, big oil, big pharma,big agriculture, weapons manufacturers, and other groups representing existing elite monied interests. The current priorities of the U.S. federal government and of globalist New World Order organizations directly fly in the face of what we must do to survive the crisis of civilization. We need a clean-up like we've never seen before and some heads are likely to roll. So be it. The world can only be thankful for getting out from under this oppression.

4. *Start over the entire systems of federal and global governance.* Yes, we can still have a Constitutional executive, legislative and judicial system. We can still have a (much smaller) military, a justice department, an energy department, a treasury, publicly funded health care, environmental protection, quality education, infrastructure and all the rest. Yes, we can formulate a transition strategy to convert institutions and manpower toward the public interest, free of vested powers. Yes, we can convert our massive military, dirty energy and aerospace capabilities toward innovation in energy, the environment, food, water, health, education and infrastructure. Yes, we can create an Earth corps to clean up the environment instead of having an aggressive Army, Navy and Air Force. We can do all this without workers having to lose their jobs. Yes, we can come to peace with the rest of the world through diplomacy and compassion. The world awaits a restoration of

good will coming from a rogue nation that has outlived its usefulness as a warmongering and fearmongering empire.

5. *Form a global green democracy* whose agenda would be almost diametrically opposed to the New World Order agenda. Representatives of all nations must come together to formulate and enforce a system that would ensure peace, sustainability and justice for all peoples, while encouraging local rule wherever possible. In no case should special interests, money or secrecy determine the agendas of these governments. At the root of this should be the principles of life, liberty, equality, justice, peace and sustainability.

6. *Fearlessly foster (suppressed) innovation such as free energy.* Any systemic change will require our utmost attention to honestly assess those new sciences and technologies that can change the world. Only these systemic changes will be able to open the door to authentic transformation of the culture. We must get beyond the rhetoric of weak policies that would only slightly mitigate the effects of global climate change and pollution. We'll have to think outside the box and get into the meat of the matter. We should quickly develop those energy sources that are truly cheap, clean, safe and decentralized, such as vacuum energy, cold fusion and advanced hydrogen technologies. No existing technologies such as solar, wind or biofuels are up to most of the task; we will need to innovate and transcend the promotions of the multitude of special interests that become vested in this or that existing technology. I urge you to consider the truth of breakthrough clean energy before you embark on dangerous climate engineering projects such as spewing particles into the atmosphere. Persuing the latest fad can only cloud our judgment and action. No existing public or private institution wants to support these hidden truths and so it will become necessary to dissolve those institutions vested in old ways and start new ones that can support rather than suppress the deeper truths and opportunities of our times. The unsung heroes of innovation will need all the help they can get to team together in an Apollo program for new energy development, frontier science and consciousness. These research and development projects will become the cornerstones of a whole new civilization that could save us from ourselves and from those of us who insist that change can only be incremental or structural.

Mr. President, you must know we all are entering the gravest crisis the world has ever experienced and that the situation can be addressed only by implementing the kinds of systemic changes listed above. Many of us are willing to support these efforts in team-work with you. I believe you will have no other choice but to move into these changes briskly. Otherwise, the degree of unrest, fear and repression will be too great to allow us to act without further violence, totalitarian control and ecological and economic collapse. We don't want that kind of world, we want to have room in which to innovate our way from the very systems that have become so decadent, so destructive, so tyrannical.

Is this an impossible task? Not if we act radically, decisively and quickly. We can only try. Crisis breeds opportunity. It is time to restore the ideals upon which our nation was founded. We have grievously lost our way from practicing those principles. We are also rapidly losing a natural environment that can nurture us all on this fragile spaceship we call Earth.

Mr. Obama, I appeal to your intelligence, wisdom and compassion to begin to facilitate the dialogue that will allow us to create those new systems that can foster the kind of future world we really want to enjoy for ourselves, our children and their children. I in no way mean my critique to be personal, I only want to help build a fire under all of us to begin the journey toward an exciting and positive new paradigm. Thank you.

— Brian O'Leary, Ph.D.
April 2009

Acknowledgements

This has been a difficult book to write, because most of the world doesn't believe what I am saying, except perhaps my description of the tyrannies that surround us and even these conclusions can be in dispute, such as 9/11 truth.

Compare the possibility of free energy with the propaganda of "solutions" such as nuclear power, biofuels, "clean coal" and carbon cap-and-trading, and we have entered a major conundrum of salesmanship. There are simply no vested interests in free energy found in the confusing cacophony of this or that publicized remedy to our energy problems. Add to all this is the decline of civilization and nature itself, the perception that no transcendental solution can possibly exist, and media access is basically zero, we have a major problem of communication.

For the past six years, this has been a lonely effort which has involved spurts of expression punctuated by thoughtful remorse about our decline, based on observations of ever-more expanding tyranny of leadership. I deeply thank my wife Meredith Miller for enduring and supporting my efforts, especially during the darkest information and feelings I have pondered and expressed. I deeply grieve the loss of Gaia and the inability of most humans to come to grips with the truth.

I also thank my new friend Keith Lampe (aka, Ponderosa Pine), the quintessential 1960s activist who writes a daily Internet newsletter, which strongly supports free energy as a solution to the deepest problems we face. Lampe is an unusual and humble balance of the four cultures that must come together to bring free energy forward.

I thank Wade Frazier for his continuing support through the years of my journey through the world of breakthrough energy. His extraordinary website www.ahealedplanet.net presents his own long journey and insights that much resemble my own, yet independently developed and grounded in history and whole systems thinking. This provides me some encouragement that I am not alone in my long journey. His generous and kindred spirit and careful reading of the manuscript are deeply appreciated.

I thank George Green and Gill DeArmond for supporting our work, housing our books and publishing them as we travel through the stresses of a new book and of opening our new retreat and gallery/bookstore. I thank George's friend Greg Caton for recently curing me of a nasty skin cancer through alternative means, saving me from going under the knife of a surgeon. I thank Jon Cypher, Joel Garbon, Joe Simonetta and several others for editing some of the essays upon this book was created.

Finally, Shaun Saunders, a brilliant writer of speculative fiction and a psychologist specializing in consumer issues as a distraction from what we should really be doing, has come forward to give this manuscript a careful look and to write the most relevant Foreword and Afterword for this book I could ever imagine.

Many others have also made my life so much more satisfying during these years of transition from a gloomy life in the U.S. to a hopeful one in Ecuador. This kind of moral support will help us play a continuing role in teaching and transforming the world from our healthy retreat Montesueños. For that, I also thank many kindred Ecuadorians for warmly supporting and welcoming us.

Appendix I
Renewable and Unconventional Energy for a Sustainable Future: Can we convert in time?

Brian O'Leary, Ph.D.*, www.brianoleary.com

Presented to the Int'l Energy Conference and Exhibition, Daigu, South Korea, Feb., 2007

Abstract

Renewable energy now comprises about 7 percent of the world's energy mix. Nuclear power consumes another 5 percent, but is radioactive, vulnerable and expensive. The remaining 88 percent comes from burning nonrenewable oil, coal and natural gas, all of which inject vast amounts of carbon dioxide into the atmosphere. Climate scientists are now making an urgent call to reduce global carbon emissions by as much as 90 percent during the coming decades. A steep increase in the use of renewables will become necessary to meet current and projected energy demands. But the existing renewables (solar, wind, biomass, hydroelectric, geothermal, etc.) each have their limitations, and so we will also need to turn towards true innovation in our energy practices and infrastructures worldwide.

Research and development on advanced, unconventional renewables has thus far been suppressed by very powerful interests. However, new policies could quickly reveal that vacuum (zero-point) energy, cold fusion, advanced hydrogen and water chemistries and other novel approaches can lead to a quantum leap in having a clean, cheap, safe, decentralized energy future. New energy technologies may be our only lasting hope to reverse the global climate crisis. We must therefore support the responsible development of these sources for a sustainable future Earth.

---

* Dr. Brian O'Leary is author of many books on science, energy and the environment. He has served on the faculties of Cornell, UC Berkeley, Hampshire and Princeton Universities and now teaches at the University of Philosophical Research. During his long and distinguished career, he has advised U.S. Congress and many presidential candidates and has worked with NASA in their scientist-astronaut and planetary programs. He is founder of the New Energy Movement. A Fellow of the AAAS and the World Innovation Foundation, he now writes, speaks and teaches all over the world and hosts environmental seminars at Montesueños, his retreat center in the Andes of Ecuador.

Introduction

Thank you for this opportunity to share some information and thoughts about renewable and innovative energy sources. It is especially an honor to have this chance for all of us here today to take an objective look at the full range of options available to us and to think deeply and honestly about what kinds of energy we would really like to have for an environmentally clean and affordable future for all of humanity.

Especially here in Northeast Asia, where you are experiencing steady economic growth and a robust intellectual and manufacturing capacity, you will be able to select from many energy options for your future. Sometimes, however, these choices are very difficult to evaluate because of the tremendous imperial, military and mega-corporate control of our global energy policies, as well as the promotional biases of those financially vested in each source. It can get very confusing when we hear of "clean coal," "sustainable biomass," "safe nuclear" and other such attributes that may not be really true in the long run. What are touted as innovative or breakthrough technologies might turn out to be only incrementally beneficial, at the very best.

Ideally, we all want to be able to use energy sources that are clean, cheap, safe, decentralized and publicly transparent—for the benefit of all humankind. But an honest assessment of these qualities requires the study of the *full life-cycle environmental and economic costs and benefits* of each option. This, then, becomes a complex job of assessing the full implications of the pervasive deployment of each option in various regions of the world. While I will only touch on the most basic aspects of new and renewable energy, this kind of deep study does need to be done for each locale in order to make the wisest choices. [1,2] No subject, besides the creation of international peace and justice, deserves a more careful look than the environmental impact of ways in which we generate and use energy. This is an issue that profoundly affects all of us and future generations. We must act soon!

What is most disconcerting to all of us here today is the alarm now being sent out by the international climate science community to drastically cut our carbon emissions. As an atmospheric and energy physicist myself, I am particularly aware of the tremendous burden posed by our collective burning of nonrenewable fossil-fuel hydrocarbons found in the ground—oil, coal and natural gas. The overwhelming consensus of thousands of climate scientists is that our collective routine use of transportation systems, power plants and heating-cooling-cooking-burning practices are now producing more greenhouse gases than we can tolerate to maintain a stable climate.

The predictions are grim. In their 2007 report [3], the scientists on the Intergovernmental Panel on Climate Change (IPCC) show that, at our current and projected rate of greenhouse gas emissions, we will melt the icecaps, glaciers and permafrost, causing significant rises in sea level. Melting the Greenland ice sheet alone could raise sea levels by several meters, inundating most coastal populations on Earth. The predicted global warming will create more and more deserts, super-storms, droughts, fires, floods, heat waves and water supply short-ages. They will deflect disease vectors northwards, create mass migrations away from the most affected areas, and will kill more and more species, including life in the ocean as it becomes ever more overfished and acidic.

Within decades, these effects will be so physically and economi-cally catastrophic, most or all of us may not be able to survive this. Even if we were to immediately cut to zero these emissions, the exist-ing imbalance of carbon dioxide in our atmosphere will go on for more than fifty years, sea levels would not go down for at least anoth-er 1000 years, so a lot of the damage is already done and will contin-ue to be done in the foreseeable future. Nobody knows for sure how soon or by how much all this will happen or whether we have already reached a "tipping point", but it would seem, at the very least, we must change our energy practices as a precautionary measure.

Many of our leading spokespeople such as Al Gore [4] and George Monbiot [5] have proposed that we shall have to reduce our carbon emissions by 70-90% or more by 2050 to have any chance to survive the positive feedback loops of runaway global warming. [6] But the reality of our times shows an *increase,* not a decrease, of our total fos-sil fuel combustion. The Kyoto Protocols to limit emissions have been ignored by some nations, sadly including the United States, yet even that modest measure appears to be too little too late to avoid a climate catastrophe. These are very real physical problems mandat-ing physical solutions which need to lie outside fossil fuels if we are to meet even our current energy demands, let alone future demands as growing economies such as those of Korea, Japan and China seek to use more energy. Hopefully this increased energy production should be free of imported oil and also create the least amount of emissions.

There are also serious economic challenges coming from the non-renewability of oil, natural gas and water. Many financial and aca-demic experts have predicted drastic price increases and conflicts because of dwindling supplies. [7] They believe we are now approach-ing peak oil and gas production, in which half the amount of total known global supply has already been used, with steadily increasing prices to come.

This perception of scarcity also creates international tensions, as the U.S. and its allies go to war simply to secure their thirst for oil. This problem can get only worse in time, unless we act soon. The American energy scientist Michael T. Klare [8] has even suggested that, as oil and other basic fuels become ever more depleted, the competition for the remaining resources is creating what he calls *petro-fascism*, a kind of totalitarian control not only of the fuels but of our very freedoms. I am embarrassed to admit that the U.S. leads the way down this destructive path. The picture is grim and so the energy choices we make now will have profound impact on our collective future.

Nuclear power in an electrified economy has also been proposed as a major energy option to help mitigate global warming. [9] But the questionable safety, vulnerability and great expense of these central station plants also raise serious questions about our collective future. The problem of the disposal of long-lived high-level radioactive waste is still not solved, exposing ourselves and untold future generations to fallout. Moreover, the proliferation of the technology poses big problems as nuclear materials find their way into the wrong hands. Anyone could, in principle, adapt these fuels to doomsday weapons, a problem I'm sure is familiar to many of you here today, just south of the 38th Parallel.

Because of dwindling uranium supplies, the possibility of making nuclear power more renewable by building breeder reactors can only increase the dangers of proliferation because the fuel produced and recycled into the reactor is highly radioactive weapons-grade plutonium.

Centralized nuclear "hot" fusion reactors are not yet proven in spite of tens of billions of dollars already spent, and are unlikely to solve the issue. These plants would have to be enormous and therefore be vulnerable to sabotage and failure. The nuclear path would also have to rely on antiquated, costly and unsightly grid systems.

Existing Renewable Energy

The only remaining choices to consider for our energy future are the renewables— hydropower, geothermal, biomass, solar thermal, solar electric, wind, tides, waves, ocean- thermal-gradients, and new, unconventional technologies such as vacuum (zero-point) energy, low-temperature nonradioactive nuclear reactions ("cold" fusion), and advanced hydrogen and water chemistries. During the rest of this presentation I will very briefly discuss each of these options for their potential in a mix of future energy choices.

Renewable energy is a bit of a misnomer. In principle, a renewable energy source is one which is always replaceable. For example, solar energy is renewable only as long as the sun shines, wind energy is renewable as long as the wind blows, hydropower is renewable as long as the water flows, biomass is renewable if we can keep growing new, dedicated crops, etc. However, a renewable source can be dirty, expensive and unsustainable when we consider its *full life-cycle* environmental impact, one in which the costs are not externalized. Therefore I prefer the term *sustainable energy* as a more important objective than renewable energy. Sustainable energy is both clean and renewable.

The existing conventional renewable energy sources are as follows:

Hydropower. Currently, about 3% of the world's energy is generated by hydropower, or falling water which turns turbines at the bottom of dams to generate electricity. If global energy demand increases, it is unlikely that hydropower's contribution to the total electricity use will be much more than it is. This is because, most of the great rivers have already been tapped for many uses, and some don't even reach their mouths. Also, the trend now is to destroy old dams rather than build new dams because of their negative environmental impact. For example, in the U.S. now, many dams are being destroyed to make way for the salmon to once again be able to swim upstream to spawn and thus avoid their own extinction.

Geothermal. Some areas have a considerable heat supply beneath the Earth, for example in Iceland. About 0.4% of the world's energy supply is geothermal. As in the case of hydropower, we are unlikely to find enough resource to be able to increase its contribution to growing global energy demand worldwide.

Solar thermal. Concentrating mirrors can focus solar energy onto pipes to heat water or for cooking or for generating electricity. Buildings with glass facing South (in Northern Hemisphere winter) can also be heated passively by sunlight. These technologies have been with us for a long time, and have many advantages, particularly in sunny areas, because of their low cost and decentralized qualities. But the share of this approach to the total global energy mix is not expected to increase much because of limited site suitabilility, land-and-materials intensity, the intermittency and diffuseness of sunlight, and the need for bulky storage systems if we want base-load (continuous) power.

Solar photovoltaics. Currently, specially-prepared silicon solar collectors can convert sunlight into direct current electricity. These cells have many of the same problems as solar thermal collectors: site sensitivity, diffuseness, intermittency, and materials-capital-and-land intensity for collection and storage. However, promising research in photovoltaic thin films could lower this cost appreciably and could make it competitive for certain local uses of non-baseload electricity. Solar-generated electricity can also be placed into a grid system so that it could supplement baseload power during daytime in certain regions. At the moment, all forms of solar energy comprise less than 0.1 % of global energy demand, but its proportion could increase depending on the choices we make for the future.

Wind power. Generating electricity from wind turbines is on the increase globally. It too is renewable yet limited, depending on the amount of wind available, and therefore has many of the same kinds of advantages and disadvantages as solar electricity. In Denmark and Germany, especially, wind power is quite cost-competitive and cleaner than burning fossil or nuclear fuels. But even there, we run into some of the same limitations posed by the other renewables: materials-and-land intensity, site sensitivity, diffuseness, intermittency and ugliness. Wind power now consumes almost 0.2% of global energy demand, and could increase somewhat, again depending on the choices we will need to make.

Biomass. Recently, there has been a push to develop ethanol from corn and other crops. As additives to gasoline, these fuels have generated marginally fewer greenhouse gases, because some of the carbon dioxide produced by burning these carbohydrates can be absorbed within a new crop, and is therefore partially renewable. However, the ethanol-from-corn option suffers from requiring a large fossil fuel infrastructure to harvest, transport and store it. Biofuels also compete with agriculture for dwindling land space. Soil can get depleted and carbon is still injected into the atmosphere. Biomass contributes about 3% of global energy demand, mostly for industrial and residential use and for ethanol and biodiesel for transportation.

Tides, waves, and ocean thermal gradient power. Each of these options, not yet developed, could provide a small fraction of the total mix. Again, we have the problems of materials intensity and water use, site suitability and diffuseness.

Hydrogen. Nowadays we hear often about the potential of hydrogen to produce electricity in fuel cells or to be burned as a fuel. But hydrogen is not an energy *source,* it is an energy *carrier.* For most applications, it takes more energy to produce hydrogen than it yields, for example, in the electrolysis of water. Hydrogen is also difficult to transport and store. On the other hand, a centralized hybrid system of solar-hydrogen energy has been proposed as one possible solution to replace fossil fuels and nuclear power to meet global energy demand.[10] Nevertheless, the grid-and-pipeline infrastructure and land-use requirements are very great, unless we can produce the hydrogen using clean unconventional energy sources such as those described below.

Summary of the traditional renewables. The above described renewable energy systems can fill some of the gaps in energy supply in some regions, but cannot fill the entire global energy demand. After years of study, I have reluctantly come to this conclusion. There simply is not enough resource, land, materials or reliability to replace fossil fuels and nuclear energy to the degree hoped for, using today's technology. Many scientists have come to the same unhappy conclusion: Richard Heinberg[11], James Lovelock[9], John Holdren and Nathan Lewis[12], for example. Lovelock has gone so far as to suggest we must return to nuclear power, because we have no other choice. Table 1 summarizes those nonrenewable and conventional renewable energy technologies which are currently known and supported by the governments and industries worldwide.

Table 1. Existing or well-researched nonrenewable and renewable energy generation technologies and their current (and projected for the foreseeable future) contribution to the global energy mix.

Nonrenewables (93%);

- Combustion of petroleum-based fuels (39%)
- Combustion of natural gas (24%)
- Combustion of coal and its derivatives (24%)
- Hydrogen derived from petroleum, natural gas, or coal
- Uranium and plutonium fission-based nuclear reactors that are highly radioactive (5%)
- "Hot fusion" (Tokamak-related) technologies supported by the U.S. Department of Energy research programs
- Renewables (7%)
- Wind-based generation systems (0.2%)
- Solar-based heating and power generation systems (0.1%)

- Geothermal-based heating and power generation systems (0.4%)
- Biofuels (ethanol and biodiesel) (1%)
- Biomass combustion (mostly wood chips) (2%)
- Fuel cells
- Tidal or wave energy electrical generators
- Thermal gradient-to-electricity processes, including ocean thermal energy conversion
- Anaerobic digestion of waste to biogas
- Conventional hydroelectric generators (3%)
- Any other technologies currently supported by existing research programs

---

Author George Monbiot [5] agrees that none of the above technologies can satisfy global energy demand and that a cut-back of 90% of our hydrocarbon energy use over the next 25 years will become necessary, requiring extreme sacrifice alongside localized improvements in efficiency and the use of renewables. He has suggested that everyone on Earth be given a carbon emissions quota that is drastically less than is now consumed by the more affluent users. In some cases there appears to be no solution, such as carbon emissions from air travel. (It shocks me to think that my trip all the way here from Ecuador and back has already used up my own quota for a lifetime!)

Therefore, in my opinion, we must now turn to the potential of innovation to solve this vexing problem, which seems to suggest that, while a degree of improvement is possible by using locally available renewables and improving the efficiency of our energy systems, no one solution is really satisfactory to deliver anything truly clean if our energy appetite were to continue even at its current level. Fortunately, such a set of solutions does appear to exist but has been suppressed for decades. It is almost heresy to imagine supplanting the world's first and only multi-trillion dollar economy vested mostly in fossil fuels. Yet this is what we might have to do to survive the climate crisis.

Unconventional Renewable Energy

The lasting solution to the energy crisis appears to be unconventional ("free") energy, which promises a quantum leap in our ability to tap clean, cheap, renewable, decentralized, safe and sustainable energy for the whole planet, thus ending the nightmares of climate change, pollution, oil wars, resource depletion, nuclear proliferation, poverty and water shortages ("free" energy could desalinate sea water).

Over the past fifteen years, I have made an intense, personal study of the potential of new energy. I traveled the world and visited some of the best and brightest scientists in the field. [2,13] Surprisingly, I have found proofs-of-concept of many technologies. With a modicum of R&D investment, new energy could be developed and deployed within the next ten years. This would be a great project for Asian governments and industries!

When full life-cycle environmental costs are considered on a global scale, none of the existing nonrenewable and renewable energy technologies appear to meet the criteria of sustainability—absent a breakthrough. By choosing selected traditional renewables in favorable areas, we could only hope for incremental changes in our energy supply in the face of accelerating global demand. More importantly, these alternatives do not address the urgent mandate for clean energy needed to mitigate global warming.

On the other hand, many new energy technologies have already been proven in hundreds of laboratories scattered throughout the world. [14] Any one or some of these approaches, if properly developed, could end our dangerous dependence on hydrocarbons and uranium. Clearly the traditional technologies keep us mired in the nineteenth and twentieth centuries rather than launching us forward into the twenty-first century. But this conventional thinking continues to dominate the news these days. Despite the great need, suppression of new energy has been historically documented in great detail by those who have taken the time to investigate. [2,13] Inventors have suffered funding cuts, threats, sabotage and even assassination ever since the time of Nicola Tesla more than one century ago.

We define "new energy" to generally mean innovative technologies with the potential of providing a quantum leap in our ability to tap cheap, clean, safe and decentralized energy for producing fuels and electricity. These may or may not be recognized by mainstream science. The technologies include:

ADVANCED HYDROGEN TECHNOLOGIES [1] catalytic water molecule manipulation and dissociation through cheap electrolysis, and [2] manipulation of hydrogen plasmas with catalysts to induce fractional quantum electronic states that yield large energy outputs;

COLD FUSION or non-radioactive low-temperature lattuce-assisted nuclear reactions by electrochemical means, induced in water and heavy water solutions catalyzed by [1] palladium cathodes,[2] sonocavitation and [3] other processes that can produce large amounts of thermal, radiation-free nuclear energy;

VACUUM ENERGY or zero-point energy, tapping the enormous quantum potential of every point in space-time, through the use of [1] super-motors with super-magnets (cf., the experiments of Michael Faraday in the 1830s),[2] solid state devices,[3] Tesla coils, and[4] charge clusters; and

THERMAL ENERGY from the environment.

Any one of the above approaches to new energy promises a quantum leap, i.e., orders of magnitude increase, in our ability to tap and have abundant clean, cheap, decentralized energy for all of humanity. In addition, there are many important transitional technologies which can mitigate emissions in the very near future, as follows:

RECYLING AND SEQUESTRATION OF CO2 AND OTHER POLLUTANTS AT THE SOURCE through innovative chemistry; and

REMEDIATION OF RADIOACTIVE NUCLEAR WASTE with innovative technologies, based on the principles of low temperature non-radioactive nuclear transmutations.

All of the above concepts have already been demonstrated in laboratories throughout the world (I have seen many such demonstrations) and have been published in the peer-reviewed literature. [14] But implementing them has proven difficult because there is no significant support for the R&D. This lack of support for unconventional thinking is familiar to those who know the history of innovation. That is to say, there is generally a bias against the credibility of a new technology until it is accepted by the mainstream culture. The most strident objectors are often scientists themselves because some of their treasured "laws" appear to be broken by breakthrough experiments that often lead to profound technological change. The bigger the change the bigger still is the resistance, by a large margin.

In spite of these severe limitations, I propose here that the transformation of our energy culture to one based on new energy may be necessary for our survival, and that we should embark on a research and development program as soon as possible. On the other hand, if the world community decides not to develop new energy [2], then we must all look at the implications of Monbiot's scenario of power-down, conventional renewables and increased conservation.[5] I summarize in Table 2 some specific new energy technologies now being researched, but currently unsupported to any significant degree:

Table 2. New and unconventional approaches to energy generation may include, but are not limited to, energy generation systems based on:

- Manipulation of electric and/or magnetic fields with novel circuits, materials, or fluids with reciprocating and/or rotating platforms
- Catalytic activation of electron energy levels in hydrogen, noble gas, or molecular gas plasma
- Zero-point energy conversion and/or Casimir effect nanotechnology engines, Van der Waals force devices, zero-bias diodes and/or non-thermal rectifiers
- Non-radioactive, aneutronic, or minimally radioactive low-temperature fission or fusion reactors
- Novel nuclear waste remediation processes
- Novel chemical carbon sequestration methods at the source

Data kindly provided by Joel Garbon

The Challenge

Political discussions about global warming and climate change have matured to the point where we all recognize we have a very big problem, and that we all must cut back on our carbon emissions. But there is not yet any serious attention paid to the solutions. The conventional renewables amount to 7 % of current and projected energy demand through the year 2030. [15] An additional small percentage can be gained by increasing the efficiency of our energy systems such as phasing in hybrid cars, better insulation and passive solar heating of buildings, and power co-generation. More savings can come from cutting down our demand for energy.

While nations such as Iceland (geothermal), Denmark (wind), Ecuador (hydropower), Brazil (ethanol from sugar cane) and desert nations (solar) are fortunate enough to potentially provide a high percentage of their energy from well-established renewables, most of the world is not so lucky. Each conventional renewable source has its limitations either because of high materials and land use, site suitability, intermittency and diffuseness—or a combination of these.

On the other hand, the new energy technologies, now unsupported, give us a chance for a breakthrough towards a global economy in which we could have clean, cheap, decentralized power and fuels. Someone, somewhere in this world will need to acknowledge and support this research and development, free of political pressures to the contrary. Just as importantly, our society may need to adapt to the potential reality of a new energy revolution.

But most importantly, we should change our political and corporate power structures now in control of our energy policies. Gone must be our carving up the Middle East for the "big prize" of oil, immortalized by a neoconservative- and energy-conglomerate-controlled secret energy task force convened in 2001 by U.S. Vice President Dick Cheney. Gone must be the assumption that we can blithely burn fossil fuels and uranium towards our own oblivion. [16] Any nation or individual who really wants to look at the most appealing options should consider a blend of both the traditional and unconventional renewable energy choices, plus placing severe limits on carbon emissions. This option can provide us with answers if we only have the courage to change.

Mother Earth is dying. She is in the Emergency Room with a high fever induced by human neglect, greed and stupidity. Do we have the will to come together in world community and create those social structures that could facilitate the transition to a clean energy future? We can only try. We and all of nature must depend on our taking responsibility to correct our grievous actions. By supporting innovation, Northeast Asia could lead the way.

References and Notes

1. *State of the World,* annual report of the Worldwatch Institute, Washington, D.C. and Norton Press, New York and London.

2. O'Leary, Brian, *Re-Inheriting the Earth,* 2003, www.brianoleary.com or P.O. Box 258, Loja, Ecuador (available in English and Spanish).

3. United Nations Intergovernmental Panel on Climate Change (IPCC), Fourth Assessment, 2007.

4. Al Gore, *An Inconvenient Truth,* winner of the Academy Award for Best Documentary of 2006.

5. George Monbiot, *Heat,* Penguin Books, London, 2006.

6. Because the recent IPCC report represents a conservatively biased consensus of climate scientists, some scientists predict much more accelerated global warming because of the IPCC's underestimate of the rate of melting of the Antarctica and Greenland ice shelves, and nonlinearities such as the accelerating melting of the Siberian permafrost, which is releasing large quantities of methane, twenty times more powerful than carbon dioxide as a greenhouse gas. Also, our oceans and forests are becoming less efficient as absorbers of carbon dioxide. Dr. James Hansen believes global warming might be passing a tipping point towards runaway warming beyond which there will be no return.

7. Many scientists and economists believe we are reaching peak production of oil in this decade, a point beyond which decreasing supplies will raise prices to levels high enough to destroy the world economy. Look under "peak oil" on the Internet for discussions of this issue.

8. Michael T. Klare, "Is Petro-Fascism in Your Future?", www.TomDispatch.com, Jan. 2007.

9. Leading environmentalist scientist, Dr. James Lovelock has advocated nuclear power as the only safe, potentially pervasive energy technology that could get us away from carbon emissions, www.ecolo.org/lovelock; "Our Nuclear Lifeline", *Readers Digest,* April 2005.

10. John O'M Bockris and T. Nejat Veziroglu, *Solar Hydrogen Energy,* Optima Press, London, 1991.

11. Richard Heinberg, *The Party's Over, Powerdown, The Oil Depletion Protocol,* New Society Publishers, 2005 and 2006.

12. Drs. John Holdren (Harvard) and Nathan Lewis (Caltech), presented to the summer 2006 Aspen Ideas Festival. For a further discussion of why environmentalists find those conventional renewables now available have problems, see Lisa Baker, "Impossible to Please", *BrainstormNW,* Lake Oswego, Oregon, December 2006.

13. Brian O'Leary, *Miracle in the Void,* Kamapua'a Press, Loja, Ecuador, 1996.

14. The website www.newenergycongress.org lists the top 100 unconventional energy technologies; www.brianoleary.com and www.ahealedplanet.net describe some rationales for supporting new energy R&D. My recent essay "Call for a New Energy Revolution" is published in *Scientific Discovery* (Jan. '07), the newsletter of the World Innovation Foundation, www.thewif.org.uk

15. Statistics from the U.S. Department of Energy's Energy Information Administration show that the projected energy mix worldwide will change very little between now and 2030, certainly by not nearly enough to address global climate change and air pollution.

16. The U.S. government has defied not only the Kyoto climate accords, it has accused the IPCC report itself as being too pessimistic, in contradiction to the assessments of several reputable climate scientists, including ones who work for the U.S. government such as Dr. James Hansen, Director of the NASA Goddard Institute for Space Studies. Instead, the Bush administration has proposed such far-out solutions as injecting particles into the atmosphere (chemtrails?) or deploying orbital mirrors that would reflect sunlight back into space, as measures to compensate for global warming. This "global dimming" would be a very expensive, polluting and dangerous action in itself, because of the uncertainties in the long-term climate change mechanisms. The U.S. Department of Energy is also funding substantial efforts to sequester carbon near coal mines, another grosser-level solution. This intention to "terror-form" the Earth would be messy, expensive and an excuse to keep emitting greenhouse gases. It seems that U.S. energy policies are geared more towards insisting we all rely on resources and technologies that enhance the profits of giant energy and military conglomerates than on searching for ways to develop clean energy and reverse global warming. Not only does the U.S. not support new energy research, but its flagship National Renewable Energy Laboratory (NERL) flounders with a steady annual $200 million budget, equivalent to one day of fighting in Iraq or two days of profits for ExxonMobil. Environmental scientist John Holdren of Harvard University agrees: "We are not starting to address climate change with the technology we have in hand, and we are not accelerating our

investment in energy technology research and development."
The Bush Administration's refusal to place any limits on carbon
emissions and its recent proposal to quintuple the use of ethanol-
from-corn to supplement gasoline will probably end up with a
policy in which we all emit *more* carbon than just burning petroleum without the ethanol.   Growing biofuels also consumes land
that could otherwise be used for agriculture. It would only enrich
big agribusiness. We must find ways to reverse these destructive
"solutions" that merely aggrandize the military-industrial complex, Wall Street corporatism and the maintenance of centralized
control of our energy technologies. The Bush administration's
lies about climate scientists' reports on climate change and its
refusal to fund any significant R&D of *any* renewables is creating
enormous suffering for all of us. These policies must be opposed.

INDEX

Praise for
The Energy Solution Revolution

"The topic and the concept behind it are timely, necessary, and needed. I congratulate you on developing a very important publication."
 – Stan McDaniel, professor emeritus of philosophy,
Sonoma State University

"*The Energy Solution Revolution* is a call for action and a guidebook for anyone who wants to positively change the course of history and save the planet."
 – Paul Von Ward, author of *Our Solarian Legacy*

"Good luck with this great new book – the world needs to move decisively into new energy and create a sustainable civilization NOW!"
 – Dr. Steven Greer, director, The Orion Project and
author of *Hidden Truth-Forbidden Knowledge*

"Your book is a tremendous and fascinating effort, and one that is very desperately needed. This is a book that real researchers will have to have in their home library permanently."
 – Tom Bearden, author of *Energy from the Vacuum*

"The issues you address in this book are about the most important ones facing humanity today. Your amicable and engaging style is consistent throughout your writings."
 – Wade Frazier, C.P.A. and author of www.ahealedplanet.net

Also by Brian O'Leary

The Making of an Ex-Astronaut

The Fertile Stars

Spaceship Titanic

Project Space Station

Mars 1999

Exploring Inner and Outer Space

The Second Coming of Science

Miracle in the Void

Re-Inheriting the Earth

9090649R0

Made in the USA
Lexington, KY
28 March 2011